GENERAL EDITOR Edward Goldsmith

PROSPECT
FOR
MAN

---

*'Quant à la destinée
humaine pessimiste,
je suis optimiste
quant à l'Homme.'—*
ALBERT CAMUS

Who will eat?

THE
WORLD
FOOD
PROBLEM
-
CAN WE
SOLVE IT?

Who will eat?

Michael Allaby

Tom Stacey

First published in 1972
by Tom Stacey Ltd.,
28–29 Maiden Lane, London WC2E 7JP

ISBN 0 85468 153 1

Printed in Great Britain
by The Garden City Press Limited
Letchworth, Hertfordshire SG6 1JS

Acknowledgements

The author wishes to acknowledge the following publications from which graphs, diagrams and tables of statistics have been reproduced in this book: *A Blueprint for Survival*, published by *The Ecologist* (graphs on pp. 39,41); *World Balance Sheet* by R. R. Doane, published by Harper & Row (statistics on pp. 52 (tilled land), 53 (land utilisation)); *State of Food and Agriculture*, published by FAO (graphs on p. 59); *The Ecological Context* by John McHale (graphs on pp. 87, 88, based on data from *The World Food Crisis* by Georg Borgstrom, published in *Futures* journal, June 1969); *Fertiliser Statistics 1970*, published by The Fertiliser Manufacturers' Association Ltd (graph on p. 117); statistics reprinted from *Man's Impact on the Global Environment* (SCEP Report) by permission of the MIT Press, Cambridge, Massachusetts (p. 134); *Small Farmlands can Yield More*, published by FAO (diagrams on pp. 230–2).

CONTENTS

Introduction

Ecology, as the term is used today, is an approach, not a discipline. It seeks to understand the behaviour of organisms, populations, societies, in fact behavioural systems in general, as indisassociable parts of larger systems (ecosystems) of which they are part, and in isolation from which they are meaningless. Looking at things in this way, provides one with a very different picture from that provided by the conventional 'expert' approach.

The 'expert' is a specialist. He devotes his life to the study of the information contained within one of the many watertight compartments into which science is at present divided. Unluckily, the world (or ecosphere) is not divided up in this way. It developed as a single process, which explains why its parts are so closely interrelated. The systems he is studying are, as a result, both influencing and being influenced by changes, falling within the compartments studied by other experts, and of which he, unfortunately, knows nothing. He is thus in no position to understand or predict changes other than those occurring in the artificial conditions of his laboratory, from which extraneous factors have been excluded, or at best in relatively closed systems, in which such conditions are most closely approximated.

Thus the expert on food production would tend to ignore the amount of power required to produce food in the way he proposes, the pollution that would be caused by the generation of such power, and the social effects of the consequent industrialization of food production.

The 'expert' approach is deficient in another respect. The discipline in which the expert is versed tends to have evolved within a specific cultural context—that obtaining in Western, industrialized society. Polanyi shows that the basic tenets of modern economic theory have little applicability outside our society. For instance, in 'primitive' societies, man does not behave in the way that will maximize his material well-being, as economists assure us he must do, but rather in the way that will bring him the maximum social prestige.

Most of the disciplines relating to human behaviour that are at present taught in our universities can also be shown to be socio-centric. It is certainly true of agriculture. Modern agricultural practice may, or may not, be appropriate to the social and ecological conditions of Western Europe; but can it be used by tribal societies living in tropical rain-forests without disrupting their social structures or reducing their soil to dust? This brings us to a further shortcoming of the 'expert' approach.

The expert tends to be provided with a very specific brief, the basis of which he is rarely in a position to question. This might be tolerable if he had a hierarchical superior, as in the case of an army officer or business executive. But this is not the case. There is no one who is both authorized and qualified to question his brief. Thus the advice offered by a pesticide expert assumes the desirability of pesticides, a very questionable assumption. Yet there is no one around (in an official capacity) to subject it to critical analysis. Worse still, the expert's brief is almost always to maximize some specific short-term benefit, regardless of consequences. This can only create disequilibria, whose correction must give rise to further disequilibria, thereby involving one in a chain-reaction in the direction of increased instability and eventual breakdown.

The stability of an agricultural system, like that of any other system, must depend on its ability to satisfy a great many, for the most part, competing environmental requirements: in this case, cultural biological and ecological ones. In order to determine how it can achieve such a compromise, and thereby provide a real and lasting solution to the problem of feeding the world population, we must, as Michael Allaby points out in his very valuable study, examine it ecologically, i.e. as part of a much wider problem, and determine a carefully integrated programme of change designed not to achieve short-term miracles, but to restore the basic conditions of stability, outside which there is no survival.

EDWARD GOLDSMITH, Editor, *The Ecologist*

1

A Quart from a Pint pot

Over the last few years, a great deal has been written about what is usually described as 'the environmental crisis.' It is suggested that man may be close to the biological and physical limitations to his expansion on this planet. There are some who believe he has exceeded these limitations already.

Whether he has or not, clearly the planet is finite. It is of a particular, invariable size and upon and immediately beneath its surface it maintains a series of life-support systems upon which we depend for our continued survival: the biosphere into which we evolved and, more recently, the minerals contained in the upper layers of the planet's crust upon which our industries are based. Since man has managed to live without processing minerals for the greatest part of his occupancy of the planet, presumably their exhaustion is a matter of no great concern so far as the survival of the species is concerned. We did without them before, there is no real reason why we cannot learn to do without them again. The life-support systems are a very different matter. We cannot live without air, without fresh water, without sunlight, without room within which to define our territory, without food. If our expansion threatens to reduce the capacity of the biosphere to provide any of these requirements, then the extinction of man as a species must be counted as a possibility.

The argument rages back and forth between those who hold that the planet has been damaged seriously, perhaps beyond repair, and that man may be doomed to extinction within the next century, and those who believe that just as technology has created certain of our problems, so it will be able to supply solutions to them and to others that may arise. If there is danger, all that is required is a redirection of technological activity and all will be well. In the middle ground there is what probably amounts to a majority of the world's biological scientists. They believe that recovery is possible, but that action is needed and that if this action is not forthcoming very soon it will be too late.

The concern of these scientists, who, because of the areas in which they work are in a position to know better than most of us what is happening to the biosphere, is summed up in a document that was presented to the Secretary-General of the United Nations, U Thant, on 11 May 1971. It has come to be known as The Menton Declaration and it was signed by 2,100 biological scientists in twenty-three countries. The presentation was made by six representatives of them: George Wald, a Nobel Laureate; Klaus Meyer-Abisch, of the Max Planck Institute, Germany; Ole Maasloe, a Danish doctor of medicine who has attended Pugwash Conferences; Jacques Monod, a French Nobel Laureate; Conrad Istock, of the University of Rochester, USA; and Laurence Slobodkin, of the State University of Stony Point, USA. The Declaration begins:

'A message to our 3.5 million neighbours on Planet Earth from 2,100 environmental scientists.

'Widely separated though we are geographically, with very different cultures, languages, attitudes, political and religious loyalties, we are united in our time by an unprecedented common danger. This danger, of a nature and magnitude never before faced by man, is born of a confluence of several phenomena. Each of them would present us with almost unmanageable problems; together they present not only the probability of vast increases in human suffering in the immediate future, but the possibility of the extinction, or virtual extinction, of human life on earth.

'As biological and other environmental scientists, we do not speak of the feasibility of particular solutions to these problems, but of our conviction that the problems exist, are global and interrelated, and that solutions can be found only if we abandon limited selfish interests to the realization of a common need.'

The problems with which the Declaration deals are environmental deterioration, the depletion of natural resources, war and population, overcrowding and hunger.

This book is concerned with hunger – with the ability of the planet to provide food for its human inhabitants. Of this problem, the Declaration says:

'The present population of Earth is estimated at three and one half billion (thousand million) people, and calculations, based on the success of present population control programmes, put it at six and a half billion (thousand million) by the year 2000. There have been some optimistic predictions that technological and natural resources can be developed to feed, clothe and house far larger populations than this. The immediate fact is, however, that as many as two-thirds of the world's present population are suffering from malnutrition and that the threat of large-scale famine is still with us despite some nutritional advances. Pollution and ecological disruption are already affecting some food sources, and frequently efforts to raise nutritional standards are themselves polluting. Moreover, population figures are misleading, since they do not take into consideration the factor of consumption. It has been estimated that a child born in the United States today will consume during his lifetime at least twenty times as much as one born in India, and contribute about fifty times as much pollution to the environment. In terms of environmental impact, therefore, the most industrialized countries are also the most densely populated.

'Man's need for space and a degree of solitude, though difficult to state in precise terms, is real and observable. We do not live by bread alone. Even if technology could produce enough synthetic food for all, overcrowding produced by ever-rising populations is likely to have disastrous social and ecological consequences.'

Most people today must be aware that there exists a world food problem. Almost daily we hear of local disasters throughout the developing countries which lead to local food shortages and it is easy to imagine that these represent the whole of the problem. Such an impression might be reinforced by the persuasive advertising of the voluntary relief agencies which play an important role in the distribution of supplies and which often dramatize their work by showing pictures of people who are hungry. This is understandable and perfectly justifiable, but it shows only a small part of a deeper, more insidious and far more dangerous situation, in which a large proportion of the world's population is chronically under- or ill-fed. Hunger and/or malnutrition face more than half the people on this planet for the whole of their lives. This is the fact and optimistic forecasts of what the world might provide must be seen in this light. Regardless of what the future may hold, at the present time the world is not being fed. There is a good deal of discussion among scientists and economists as to what might be the maximum possible world population. Populations are growing at a rate which leads to a doubling every 35 years. For how long will it be possible to contain this growth rate? How many people can the world support? Debates such as these are at best academic. More relevant are those of others, who attempt to calculate whether the world can sustain the population it has now and, if not, by how many its carrying capacity has been exceeded.

The world shortage of food cannot be seen realistically outside of the wider context of the relationship between human population and resources, for without some of these resources there can be no food. We need water to grow plants as well as to drink. We need petroleum to transport food and to manufacture pesticides, as well as to fuel fighter planes. We need power to manufacture fertilizers as well as to operate washing machines. The list is endless and it demonstrates the interrelationship of resources.

The consumption of resources itself is related to our desire for economic and industrial growth, which we tend to equate with a belief in the inevitability of 'progress' – which we seldom attempt to define. 'You can't stop progress' is the argument most frequently heard by those who challenge the right of industry and technology to reduce the quality of life. So we learn to put up with pollution, with noise, with crowded cities, with crime, and believe that these are prices we must be willing to pay for benefits that are much harder to identify.

Perhaps 'progress' began with an idea that is basic to western philosophy: man is apart from nature. Nature exists to serve man, God's crowning achievement. Therefore, man must tame nature, must master it, dominate it, bend it to his will. The yardstick by which his civilization is measured measures, in fact, the width of the gulf he has created between himself and nature. For this reason it is difficult for him to accept that peoples living much more naturally, or to a different scale of values, such as hunter-gatherer communities, are civilized at all. Such people are 'savages' by definition.

However, since our mission to conquer nature is divine, the lot of such unfortunates must be improved. They must be enabled to see the depths of their depravity and, in their enlightenment, they will join us in creating the perfect world in which, one day, all men will live in large cities and the wicked outside will be kept outside for good.

In pursuit of this mastery of nature we have developed industries to manufacture goods to satisfy the material wants that we have generated and to regulate the artificial environments we have built for ourselves. In so doing we have increased greatly our material possessions and since it is in these terms that we measure wealth, we have become wealthy. We have also become powerful, so that we have been able to achieve a degree of cultural dominance over the entire world that is quite unprecedented. So, throughout the world, we have imposed our own values, our own definitions of wealth and of 'the good life.' Since our accumulation of material possessions has been achieved, in part, by the unfair exploitation of the people of what we now call 'developing' countries – we mean poor countries – and because of our evangelical fervour, we have compelled all those with whom we have come into contact to join in the pursuit of material things. At the same time we have disrupted their social and cultural organization, so weakening them. Now we are all involved in The Great Myth.

The Great Myth is another deep-rooted part of our culture. We believe that by working hard we will prosper. Hard work is good, is virtuous, in a moral sense. We believe that it has been our industriousness that has led to our present good fortune and that any who would enjoy it with us must follow in our footsteps. It is likely that a component of guilt can be identified in this feeling. It was not only hard work which created our material wealth, it was also the natural resources taken from other, weaker nations without payment and without their permission. Even the work we love was not ours alone: at the beginning we had slaves and then working populations we treated as slaves. Nevertheless, the Myth is pervasive. We offer it to the peoples of the poor countries. We tell them that if they work hard they can live like Californians. It is a convenient myth, for we can use it to deny them the aid they must have. 'Nothing is worth having unless you earned it for yourself,' we say, and, 'To give the idle poor money is to encourage their idleness.' We use all the arguments that were used by nineteenth-century industrialists to deny all but the barest necessities of life to their employees. We recognize the injustices in their historical context, but we overlook them when they are used between nations. Yet there is no difference, and just as a redistribution of wealth within the national community was a matter of justice and, eventually, of social survival, so it is within the world community.

It was not possible that all working men, women and children could become as wealthy as their employers; there was only so much wealth to be shared. No more is it possible for the poor countries to achieve a level of material prosperity equal to ours. There is only so much wealth in the world and we have more than our share. There is not the remotest possibility that anywhere except in California will people live as Californians. The Great Myth amounts to one of the most monumental pieces of international self-deception in history. It is a lie, there is no foundation of fact in it, and it is highly dangerous. Today it threatens our survival.

The imposition of our culture has introduced throughout the world techniques for preventing death, while at the same time it has destroyed cultural controls that once regulated population levels. This has led to the present exponential growth of population, which places upon the environment demands that are not being met. At the same time our obsession with material possessions adds to these demands a demand for industrialization – for economic growth – and this cannot be met either, for the planet lacks the resources, particularly the minerals and the water, to permit it.

So we see that the problem of feeding the world cannot be divorced from our ideas about population growth, industrial growth and materialism. Immediately, it is much more a political, social and economic problem than one concerning agriculture alone. Basically, though, it is a philosophical problem and it cannot be solved satisfactorily unless we are able and willing to re-examine and modify the way in which we see the world we live in. We must learn that it is impossible to put a quart into a pint pot – or to take a quart from one.

For this reason, the problem concerns us all. It is true that the solutions will be applied by governments, but if they involve major changes in the outlook and aims of our society, then we are involved. It is the outlook and aims of each one of us that must change. The philosophy is ours, the politicians we elect are chosen because they share it with us and seek to implement it. It is not their invention. So, if we are to have governments capable of seeing into the nature of the problem and of solving it, then we ourselves must be able to solve it for and within ourselves and elect the politicians that share our new outlook. Environmentalists in America have produced a slogan: 'If you are not part of the solution, you are part of the problem.' It is very true.

So far, most attempts to increase the amount of food available in the developing countries have been based on increasing agricultural output. The organization most involved in these attempts is the Food and Agriculture Organization of the United Nations, and the programme it has prepared and which it is trying to introduce has been given the nickname of 'The Green Revolution.' Obviously, I have had to devote a considerable amount of attention to the 'Green Revolution,' but first I have tried to define the problem as one of growth versus resources, examining the nature and implications of exponential growth rates and rates of consumption, as they affect population and resources.

Because it is somewhat technical, I have outlined the 'Green Revolution' programme in some detail. The information in Chapter 3 is taken mainly from publications issued by the FAO. It is largely, therefore, the world food problem as it is seen by the FAO.

At the centre of the 'Green Revolution,' and essential to its success, is a series of new cereal varieties, which are more responsive than traditional varieties to inputs of fertilizer and water, but which require also high inputs of pesticides. Since the availability of each of these inputs will determine the success or failure of the programme as a whole, I have devoted a chapter to each of them.

As I hope I have shown already, it is a characteristic of our relationship with the poorer countries that we encourage them to follow in our footsteps in the direction of industrialization. This applies, too, to the production of food. Over the last thirty years the developed countries have achieved a degree of industrialization of their agriculture. This has meant the introduction of new techniques and whole new farming patterns that have replaced traditional farming almost completely. Broadly, this is what we propose for the developing countries. It is what we mean when we talk of intensifying production on existing farms. How successful has it been in the developed countries? The new farming has been in business for long enough for an appraisal to be made of the results and for us to speculate about future trends. It is appropriate to take British farming as an example, because in Britain farming today is more intensive and more highly industrialized than in most countries. It is relevant for three reasons. If we are to introduce these techniques in other countries, we ought to be certain that they work at home. No less important is the fact that we, too, are part of the world food problem. We have to be fed like anyone else. If our methods of producing food are unsatisfactory then we should change them. Then, if the demand for food in the developing countries increases beyond a certain point, patterns of world trade will be affected. The developing countries will be less able to export food and, therefore, we will be less able to import it. There is a probability that we will need to become far more self-sufficient. Can we achieve self-sufficiency with our present population, with our present rate of population growth, and with our present farming methods? If the answer to any of these questions is no, then we are about to join the ranks of the hungry.

Industrial growth, urbanization and the intensification of agriculture, all increase environmental pollution. If this is the direction in which we urge developing countries to head, then we must consider the possible environmental side-effects. If, as some biologists infer, the biosphere is badly damaged already as a result of the industrial activities of the rich countries, what may be the result of extending industrialization to new areas? It is better to prepare ourselves now for difficulties we may be about to create than to attempt to solve them after we have created them, when, in any case, it may be too late.

Then we must consider the degree of industrialization of agriculture that is practicable and relevant. A large part of the increased production of the developed countries over the last few years is attributable to the intensification of livestock husbandry in indoor lots – 'factory farms.' 'Factory farming' generates heated controversy, but to what extent is it necessary to the developed countries, what does it contribute, and can it be sustained? Already it faces grave economic and biological difficulties. What is its future? (Has it a future?) Its apologists argue that because it is 'efficient' it must be tolerated if we are to feed the world's starving millions. This brings it within the discussion of the world food problem. Can it be adapted to conditions in developing countries and what is its relevance to world food production as a whole?

From this I have gone on to examine, briefly, other more fanciful or exotic solutions that have been proposed and, at the same time, to examine the contribution made by the world's fisheries. Again, it is suggested that the oceans may yield harvests undreamed of – a suggestion reminiscent of our talk of 'conquering nature.' Is it true? How much fish is there and who is eating it? The manufacture of synthetic proteins has also been advanced as a solution, either for use directly as human food or as supplementary feed for animals. Will we all be eating 'knitted steaks' in the years to come? All these enquiries lead, inevitably, to the final question. Supposing our food requirement were the only limitation on our expansion, how many people could the planet feed? This figure can be calculated in terms of ecological energetics and I have tried to explain some of the steps by which it is done.

What then? We are talking now of survival. If we had a world carrying the maximum human population, if man were pressing hard against the final limit to his expansion, what would survival be like? We can amuse ourselves by imagining life under such circumstances. Then we may be in a better position to decide whether such a life would be tolerable. If it is not, then we owe it to our successors to solve the dilemma now, rather than leaving it for them.

In fact, the availability of food could be increased considerably if waste were reduced. Food is wasted in distribution and, no less serious, human beings are wasted – the rural unemployed who could be engaged in production. To reduce rural unemployment is an important aim of the FAO and one of the central themes of the 'Green Revolution.' Yet in many of the developing countries it requires reforms on a scale so large that they amount to revolution in a political rather than an agricultural sense.

What solutions are there then? As I suggested earlier, a final solution can be achieved only by modifying our outlook on the world around us. If we are able to make this modification there is a good chance that we may escape.

The book may appear pessimistic. I do not believe that it contains any distortion of the problem, but I have not accepted uncritically the optimistic projections of the 'technophiles' – those who believe in the omnipotence of technology to rescue mankind from its own entanglements. Since the developed, and often the developing, countries are enamoured of their technologies at the present time, my scepticism will seem heretical to some. Without the technological '*deus ex machina*' there are few solutions open to us, and after I have rejected any that are based on continued economic and industrial expansion, because further expansion is not one of our options, the picture does begin to look dark. I maintain that this is not my pessimism. If the impression I give is gloomy, it is because the outlook is gloomy.

Major world famines, accompanied by war and epidemic disease, are a very real possibility before 1980. Famine in the developed countries is a very real possibility before the end of the century. Once we examine the whole of man's relationship to his global environment, not just one part of it, this is the pattern that emerges. Of course we can maximise one side of the equation if we are allowed to maximise the other. Of course we can produce infinite food so long as we assume the existence of infinite land, infinite water, infinite power. The truth is that our resources are not infinite.

It has become fashionable to talk of 'Spaceship Earth.' The comparison between the planet and a spaceship is accurate, valid and useful. We are accustomed to watching television pictures of men travelling in small capsules in a totally hostile environment. We can understand that inside their capsules they must have air, food, water and the means to remove and de-toxify their wastes. We can understand too, that if these should fail the men will die. We can understand that were they to venture from their capsule without protection they would die instantly.

The space through which they travel is the same space as that through which the planet travels. The life-support systems installed on their craft are no more than simulations of those that exist on Earth. Just as a failure on the spaceship would cause the death of its crew, so a failure on Earth would cause the death of its crew (if it has one) and passengers. The difference is only one of scale.

We can understand that if six men were to crowd into a capsule designed for three, there is a distinct probability that all six of them would die. Why do we imagine that the planet cannot be similarly overloaded, with similar results?

There is no natural law that says everything will come out all right in the end.

2

The growth of Growth

It was Thomas Malthus who first suggested that if the availability of natural resources should increase, then population will increase in proportion until it has outstripped the resources. He was talking particularly of food and the first implication of his thesis was that at all times there is bound to be a minority of people living at or below subsistence level. The poor, in fact, will always be with us. This, he argued, is a natural law and we must accept it.

Many of his contemporaries refused to accept it. The essay, published in 1798, in which he propounded his ideas provoked a lively controversy among scientists. Darwin was influenced by it when he was formulating his own theory of survival. It provoked an even livelier debate among political thinkers. Those on the left rejected it out of hand on the grounds that it was worse than wrong, it was dangerous and immoral.

It is easy to see why. If Malthus were right, then it would seem natural to many nineteenth-century industrialists to argue that it would be pointless to reform conditions of employment and increase wages for their workers. No matter how far such reforms were to be carried, the poor would still be there. There would still be starvation for some. Indeed, raising wages would stimulate demands and so accelerate the process. The best course would be to leave things as they were, to avoid interfering with the subtle laws of economics and to allow the strongest, or fittest, to survive. Seen in this way, Malthus might be said to have provided some kind of scientific credibility for the doctrine of *laissez-faire*. Obviously, the left could not accept this. Even today, the name of Malthus is anathema to many Socialists.

As the years went by it appeared that the Socialists had been right and Malthus wrong. The population did increase, but so did the general level of prosperity. Reforms were made, incomes increased and with the added spending power of the working population the economy prospered and so, eventually, did they. It is only in recent years that Malthus has emerged again from obscurity and a new breed of 'Neo-Malthusians' is warning once again of the perils of a population growth rate that is in danger of outstripping resources. As before, they attract ridicule. Could they be right? We hear a great deal about the 'population explosion,' but at present the world population is increasing at only 1.9 per cent a year. This sounds little enough. Why all the fuss?

The reason for the fuss, and the basis of the Malthusian argument, is that the growth rate is exponential. Malthus reasoned that population is bound to increase exponentially, or geometrically, while resource availability – he was speaking mainly of food, remember – is bound to increase arithmetically. What does this mean? The difference is that between simple and compound interest. If you invest £100 at a rate of 10 per cent simple interest, then your capital will increase by £10 per year. If the interest rate is 10 per cent compound, then it will increase by £10 after one year, £11 after the second year, and so on. This will alter dramatically the time taken to double the original capital – ten years in the case of simple interest, seven-and-a-half years at compound interest. Since the rate of increase is proportional to the original sum, the larger that sum the more rapidly will it increase.

The rate of increase of large numbers can be surprisingly rapid when the growth rate is exponential. There is a legend about the man who taught the king of a Middle Eastern country to play chess. The king was so delighted with the game that he offered the visitor any reward he cared to name. The wily chess master demanded one grain of wheat for the first square on the board, two for the second, four for the third, and so on. The king agreed, but when his chamberlain came to work out the amount of grain to be paid it was found that the entire wheat stocks of the country would be exhausted long before the 64th square was reached. This was, admittedly, a high rate of compound interest!

An exponential increase, then, is one in which the rate of increase is constant over time as a percentage *of the whole*. It can be drawn as a graph, in which the growth curve rises ever more steeply with time, until it is vertical. It can also be expressed as an equation.

It is simple to use the equation to calculate the doubling time of any quantity whose growth is exponential, provided the rate of growth is known. It can be applied to populations, when it goes like this:

Annual growth rate per cent	*Doubling time* years
1.0	70
2.0	35
3.0	24
4.0	17

So we see that the present world population growth rate of 1.9 per cent per year gives a doubling time of about thirty-seven years. So long as this rate is maintained, the world population will double every thirty-seven years. By the time a baby born today reaches the age of seventy-four, there will be four people in the world for every one alive now.

What about resource availability then? We may increase food production by making two blades of grass grow where one grew before, but although for a time such a growth rate may appear to be exponential, a limit to expansion will be reached. The availability of the resource will be contained within a definite limit and over time this will give an effect similar to an arithmetical, or simple interest, growth.

How was it, then, that Malthus' prediction was believed to be false? How was it that populations did continue to increase – exponentially, though at a smaller rate than that which he predicted – without outrunning food supplies? The answer is based on a non-repeatable historical accident. During the nineteenth century the European colonial powers were expanding and colonizing into North America, Australasia and Africa. All of these continents had vast natural resources which the Europeans exploited. As the colonists began to produce food, particularly in North America and later in Australia and New Zealand, they found their natural market in the homelands of Europe. In effect there was a sudden and major increase in the area of agricultural land from which food could be produced and it was this that permitted European populations to continue to grow. It was as though Europe itself had expanded to give more space and, therefore, more food. There was room for more people, even if most of them did live in urbanized Europe. Those who pronounced Malthus wrong spoke too soon. At worst his thesis must remain unproven for the time being. However, had there not been that colonial expansion in the eighteenth and nineteenth centuries it is doubtful whether Britain would be able to sustain today more than about half of its actual population – the number it can feed from home-grown food. The accident cannot happen again. In this, at least, it is not possible that history will repeat itself. There are no more Americas, Africas or Australasias.

There are other planets, of course, and there are some optimists who have suggested that the miracle might be made to repeat itself by establishing extra-terrestrial colonies that would export (what?) to Earth. Chapter 10 is devoted to 'wishful thinking' and we will return to the would-be astronauts. For the moment it is enough to say that it is no more than fantasy and in view of the economic and resource problems that will face the industrial countries in the immediate future it will remain fantasy. It has been suggested that surplus population could be exported to other planets. This does nothing to solve the problem of exponential growth. Supposing that it were technologically possible to export people to all the planets in the solar system – travelling outside the solar system will not be feasible within the foreseeable future – it would take something like 200 years to fill them to capacity if the present growth rate is maintained. The United States, as the world's richest country in terms of accumulated capital, might consider exporting its own surplus population. To do so it would have to launch 2,000 Apollo-sized spaceships every day, at a daily cost of 300,000 million dollars, which is about one-third of the country's *annual* Gross National Product!

The population explosion has happened already; it is not something that awaits us. We must accept that unless major disasters intervene to effect a massive reduction in population, by the year 2000 there will be between 6 and 7 thousand million people on this planet.

The origins of the present growth rate reach back a long way in time. It may have taken 1 million years for the world population to grow from 2½ to 5 million, the figure it reached some time around 6000 BC; 7,650 years later, in 1650, the figure reached 500 million. At that time it was increasing at about 0.3 per cent a year. This would have given it a doubling time of approximately 250 years and by the year 2000 it would have reached around 1,500 million. However, the rate of growth itself has been increasing. By 1900 it was 0.6 per cent per year (doubling time 115 years) and by 1950 it was 0.9 per cent per year (doubling time seventy-seven years).

Population growth rate is simply the birth rate minus the death rate. The birth rate has remained fairly constant, but the death rate has been declining steadily since the seventeenth century. Medical discoveries have rendered curable diseases and injuries that once were fatal. New food production techniques have provided more food but, more significantly perhaps, industrial development and the urbanization of large populations, coupled with a general increase in material wealth, has led to improvements in food processing, storage and distribution. In the industrial countries there is little food wasted and the mechanisms by which it reaches the urban population are complex and efficient. In the developing countries this is not so. There, food is not always distributed efficiently. Nevertheless, during this century in particular, there has been a spectacular drop in death rates.

It is not only a reduction in the crude death rate that has led to growing populations. What is more important is that most of the lives that have been saved have been young lives. Many of the diseases that have been brought under at least partial control used to kill more children than adults. Thus, on the one hand, life expectancy, which is calculated by dividing the number of deaths into the sum of the ages at death, has increased throughout the world. In 1650, when the death rate was only 0.3 per cent less than the birth rate, life expectancy was thirty years. In the world as a whole it is now fifty-three years. Almost all of this increase has been achieved by eliminating diseases that killed children. The life expectancy of a middle-aged Englishman has increased by only three years in the last century and the life expectancy of a middle-aged American is actually a little lower than it was some years ago. People are not

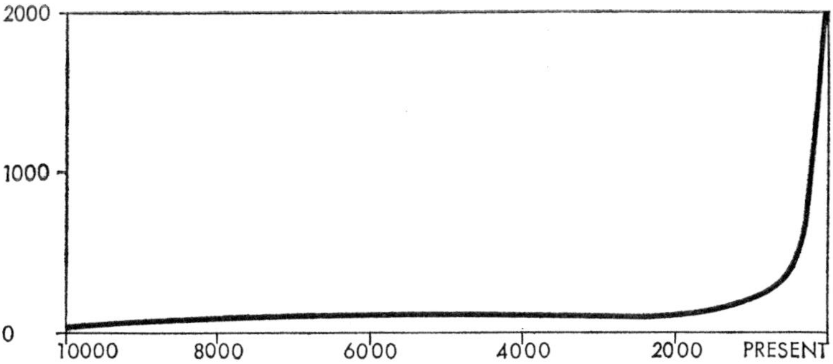

Arithmetic population curve. World population began to increase rapidly about 500 years ago.

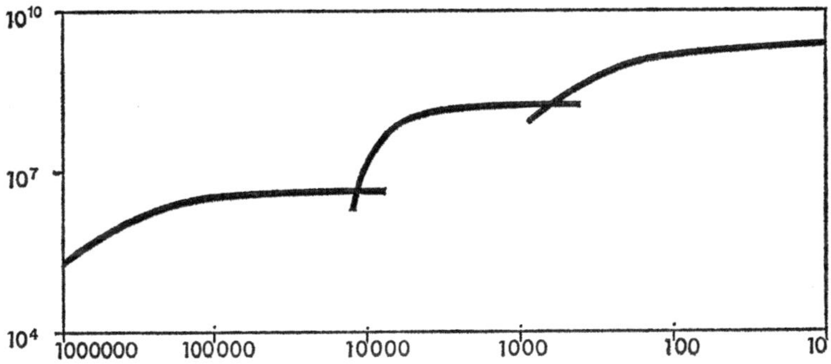

Logarithmic population curve. By using a logarithmic graph it is possible to plot population growth over a longer period of time and over a wider range. There have been three surges related to three 'revolutions': tool-making, or cultural, agricultural, scientific-industrial.

living to a greater age, it is just that more of them live their full term.

On the other hand, this also means that children that once would have died now live into the reproductive phase of their lives. More people surviving to maturity means more parents. The population is not growing older, as we might expect with an increase in life expectancy, it is growing younger. In Costa Rica, where back in 1966 the population growth rate gave a doubling time of seventeen years, more than half the total population is now under fifteen years of age. In the developing countries generally, some 40 per cent of the population is under fifteen. During the next few years this mass of people will begin to reproduce and it is reasonable to suppose that the world population growth rate, which has been tending to increase for several centuries, will continue to increase. Is the rate of increase itself subject to exponential increase?

Population growth in the developed countries, where family planning information and supplies are more readily obtainable and where, possibly, the opportunity for material advancement is reduced for those with large families, so providing an economic incentive favouring smaller families, is less of a problem. That is not to say there is no problem, nor is it to say there is no need for positive policies to check population growth. It was in 1950 that the Royal Commission on Population recommended that Britain's population should be stabilized. So far nothing has happened and no British government has grasped the nettle. The Royal Commission recommended that an official body should be made responsible for keeping a close watch on population trends. *Population*, published by the Labour Party for its 1971 women's conference, suggests that the developed countries should be prepared to accept a reduction in their standard of living in order to assist the developing countries. At the same time it pointed out that the population problem affects Britain as well as the poorer countries. (The Menton Declaration draws attention to the relative consumption of resources and environmental impact of people in developed and developing countries.) It criticized those local authorities that do not take responsibility for family planning facilities and it suggests there should be a population policy for Britain.

At last it seems at least possible that some small steps may be taken. A report of the Commons Select Committee on Science and Technology, published in May 1971, suggests the establishment of a 'special office' responsible directly to the Prime Minister. Its task would be to improve and co-ordinate studies of British and world population trends; study the relationships between population, resource availability, economic growth and the effects on the environment; advise the Government on such steps as it felt were necessary; and publicise the consequences of excessive population growth. Some of the press reports suggested some kind of 'Ministry of Population' might be established. It may come to that, but it has not done so yet; the tenor of the report is mild. The United Nations, through several of its agencies, maintains figures of world population trends and the FAO keeps a close watch on the relationship between population and food supply and availability. Monitoring data hardly amounts to a policy, but at least the matter is receiving serious attention.

What is a population policy, anyway? The decision whether or not to have children is intensely personal. To legislate in this area is bound to be distasteful and to attempt to limit family size by law might be impossible. If wicked parents transgressed, what sanctions could be imposed? The harm would have been done and to penalize the parents would be likely to harm the children more. The only rational solution would be to kill the unwanted, or at any rate statistically superfluous child. It is difficult to imagine any government enacting legislation of this kind, although infanticide and abortion have been the most common methods of family planning in many cultures throughout most of history.

Perhaps we should allow the growth to continue and try to live with it? This would be by far the most comfortable solution. It is always easiest to do nothing. We will consider shortly what the implications might be.

Is Britain over-populated? According to the Report of the Select Committee, the 1971 population of 55.7 million may be expected to increase to 62 million by 1990, 66 million by 2000. These estimates accord with those of most other authorities, who predict a population for Britain of between 60 and 70 million by the end of the century.

Is this too many? How many is enough? In 1969 the Institute of Biology, a scientific society, held a symposium in London called 'The Optimum Population for Britain.'* Delegates listened to speaker after speaker discuss population, growth rates, and the problems associated with control measures. Suggestions from the floor included one for licensing would-be parents. At the end the delegates took a vote on whether, in their view, the optimum population for Britain had been exceeded already. The audience consisted mainly

* Published as *The Optimum Population for Britain*, edited by L. R. Taylor, Academic Press, 1970.

of professional biologists with their invited guests. Ninety per cent of them registered their opinion that the optimum population had been exceeded. In the words of L. R. Taylor, convenor of the symposium, who edited the published proceedings:

'This may not represent a general public view, but it is that of a responsible group of people, disciplined in the study of living organisms, whose prime concern is the continuation of life processes. *Their opinion is not lightly to be dismissed.*'

It is not simple to decide what the optimum population for any area should be, but if the likely limiting factor is food availability, then it would seem logical to suppose that in the long term no area can support a population larger than that which it can feed from its own production. At a time when world population is increasing rapidly, it is likely that there will be restraints on international trade in food as the requirements for domestic markets grow. It may be, therefore, that Britain's current rôle as a manufacturing country selling industrial products in order to buy food may have to change. There may be less food to buy. In the end it is food that counts: we cannot eat machine tools, no matter how many of them we make. If this is so, and if the optimum population for Britain is one which can be fed from the agricultural production of Britain itself, then since we import approximately half of the temperate foodstuffs we consume, it would suggest that the optimum population should be no more than half the present population – 28 million people. The factor that could change this would be an increase or a reduction in food production in Britain. As it is, production in some major commodities has levelled off, while in others increases are now slow and minimal. The years of the dramatic increases in farm yields are over. (This is a subject to which we will return in a later chapter.) Short of some major breakthrough in plant breeding, or a significant improvement in the fertility of our soils we can expect no further major increases in yields. It is more likely, in fact, that current levels of production may decline so long as present techniques remain in use. The likelihood is that the figure of 28 million as an optimum population for Britain is on the high side and that a more realistic figure would be rather lower.

In any discussion of population and the pressure of numbers on resources in Britain, immigration is bound to be raised. It is a red herring. The balance between immigrants and emigrants shows that each year more people leave Britain than come to it. Immigrants contribute nothing to the problem. Nor do they contribute to its solution. If it were agreed that the country is over-populated, then a quick solution might be to offer high incentives to British citizens to emigrate to parts of the world that are under-populated. In any case, if the aim is to reduce the present population, then a strong case can be argued for halting all immigration, from any region, under almost any circumstances. Such a policy would take no account of country of origin, age, health or skill.

The problems that affect Britain affect other industrial countries, and if Britain is over-populated, so are they, although there are few that can compete with Britain's population density of 588 persons per square mile. (Taken separately from the remainder of the United Kingdom, England has a population density of 911 per sq. mile.) In Europe, apart from Gibraltar (12,500 per sq. m.), Malta and Gozo (2,607 per sq. m.) and Monaco, which accommodates 24,000 people in half a sq. mile, only Belgium (814 per sq. m.), West Germany (630 per sq. m.), the Netherlands (949 per sq. m.) and San Marino (783 per sq. m.) exceed the British population density, and apart from the Netherlands, England is the most densely populated country in Europe. Outside Europe there are other small areas with very high populations – Hong Kong manages somehow to squeeze nearly 4 million people into its 398 sq. miles, but, in contrast, Outer Mongolia has a rarified 2 persons per sq. mile.

In the United States (population density 57 per sq. m.) the problem is taken seriously and a voluntary organization called Zero Population Growth (ZPG) looks like achieving some success with its policy of urging the country to adopt a population policy. A coalition has been formed to back a resolution by thirty-three senators and twenty-five representatives that would have that effect. Led by Dr. Milton Eisenhower, president of Johns Hopkins University and the brother of the late President, and Joseph Tydings, a former senator, the coalition has declared that its resolution will aim for population stabilization by voluntary means. It would make birth control advice and facilities more widely available and would permit federal funds to be used for an educational programme to encourage a reduction in the average number of children in each family from the present 2.6 to 2.1. At the same time it would initiate tax reforms. At present, says the coalition, the tax structure favours large families and penalizes childless couples and single persons. The coalition has wide and diverse backing from, among others, the presidents of the United Auto Workers' Union, the National Association for the Advancement of Coloured Peoples, PanAm and the Council of the United Methodist Church.

A population policy does not mean necessarily reducing population. There are countries still which are under-populated and there are those in which a decline in population is a cause for concern. Frequently this is caused by emigration. Ireland (109 per sq. m.) is said to be one country that exports people. Portugal (266 per sq. m.) is another. A census conducted in 1960 showed a total population of 8.8 million. Portugal's ambition is to bring the population up to the 10 million mark but the preliminary results from the latest census show that there is a 13 per cent decline in population in nine of the country's twenty-two administrative districts. In one town the population has been reduced by 73 per cent. The emigrants leave for more lucrative jobs, usually in countries not too far away. There are said to be 600,000 Portuguese living in France alone. A population well below the optimum can be nearly as serious as one above it. The low numbers produce shortages of labour which hinder developments that would improve the standard of living of those that remain. So living standards stay low and more workers leave and the situation perpetuates itself.

Population growth can be checked if this is thought to be desirable. In Japan in 1947 the birth rate was 34.3 per thousand. By 1957 it had been halved to 17.2 per thousand and it has remained stable for the last fifteen years. In the early 1950s the population was increasing at 2.9 per cent per year. Now it increases at 1.0 per cent per year. According to a report in *The Times* (4 April 1971) by Michael Hornsby, 'The straitened economic circumstances of the Japanese after the war, coupled with the often lurid accounts in the news media of the fate awaiting them should population growth continue unchecked, seem to have been responsible for the characteristically disciplined way in which they set about reducing the birth rate. Induced abortion was easily the most widely practised form of birth control. The Government, somewhat in retard of the act, enacted

* *

the so-called eugenic protection law in 1949, making abortion legally obtainable for health and economic reasons and removing obstacles to sterilization and other types of birth control.' Abortions account for about 70 per cent of the decline in the birth rate and the abortion rate is now around 750,000 per year. In the late 1950s it was around one million.

Such a sudden reduction in birth rate produces imbalances in the age distribution of the population. Economically, this is reflected in the ratio of working to non-working, or dependent, population. At first the reduction in babies and small children will mean that there are relatively more workers to dependents, so the average income will rise. Soon, however, there will be an increase in the number of old people in comparison with the young and the balance will swing the other way. Average incomes will fall. This worries governments, but it is a temporary phenomenon, for as the old die, producing a temporarily high death rate, the balance will be restored. Even in the short term the risks may have been over-dramatized, since the increase in the number of old people will be partly compensated for by the reduction in the number of children. Once the population is stabilized and births once more balance deaths, the age distribution will even itself. The alternative, which is to allow the high birth rate to continue, will produce a similar effect at the other end of the range. There will be a higher number of non-productive children to support. It is arguable that this causes more hardship, on the grounds that young couples just establishing themselves in new homes and careers may be poorer than the established middle-aged people who are most likely to find themselves supporting their parents and, at the same time, schooling and welfare services for the young may place a heavier burden on the community as a whole than caring for the aged. Any change in the size of the population, either an increase or a decrease, will have an economic effect, but this will continue only until the population stabilizes itself again at a new level.

Although an unchecked exponential growth of population must be a source of concern anywhere, the immediate problem lies in the developing countries, for although the demand for resources is higher in the industrial countries, it is in the developing countries that populations are pressing hardest against the resources that are available to them at the present time. It is in these countries that the growth is fastest.

REGION	*Population growth rate*	percentage per year
Latin America		2.9
Far East		2.4
Near East		2.5
Africa		2.4

These growth rates give doubling times throughout the third world of something like thirty years. Throughout the world, individual scientists as well as the FAO, are expressing concern over the current availability of food in the developing countries and the prospects for the future should population growth not be checked. The Faculty Board of Agriculture of the University of Cambridge held a series of lectures on population and food supply in 1966 and repeated them in 1967. In his conclusion to the published proceedings,* the editor, Sir Joseph Hutchinson, emeritus Drapers' Professor of Agriculture at the University of Cambridge, wrote:

'Population growth at anything like the current rate cannot be supported for many generations. This is clear. But despite the constraints and rigidities of the current agricultural system . . . there are opportunities now within our reach to increase the world's food supply very substantially. There can be no disagreement with the view that our first obligation is to provide the food and other biological resources for all those already born, or whose arrival we can foresee. In the longer term, however, it should be accepted that agricultural production cannot be multiplied indefinitely. It is no more than common prudence to plan for the stabilization of human populations before the point is reached that food production can no longer keep pace with human multiplication, and readjustment by catastrophe becomes inevitable.'

When he wrote of the opportunities to increase world food production, Sir Joseph probably had in mind the 'Green Revolution.' We will consider this plan in more detail later, but at this stage it is

* *Population and Food Supply*, edited by Sir Joseph Hutchinson, Cambridge University Press, 1969.

only fair to point out that Sir Joseph was writing in 1969. More information is available now of the effects of the 'Green Revolution.'

One of the 'Green Revolution's' architects is Dr. Norman E. Borlaug. He was awarded the 1970 Nobel Peace Prize for his contribution to the development of the new high-yielding cereal varieties on which the programme is based. Yet in a message to the FAO's 25th Anniversary Conference, held in Rome in November 1970, he warned of what he called 'the monster' of population growth which, he said, could destroy the world. He went on:

'I am convinced that those of us working with food and agriculture also must bring pressure on politicians, government officials and the general public to face up to and tame the monster of population growth before it is too late.'

However, food is only one resource we need to consider. If food production is to increase it will require an increase in the inputs of capital and the products – agrochemicals in particular – of industry. Industry, too, is dependent on resources and just as population growth is proceeding exponentially, so is industrial growth.

We are used to thinking of the Industrial Revolution as an eighteenth and nineteenth century development and it may come as a surprise to realize that compared with the expansion of industry throughout the world as a whole since 1945, all that happened industrially before 1945 dwindles to insignificance. It has been estimated that if, somehow, all the industrial effort expended in the whole world during the whole of our history up to 1945 were to be lost, at the present rate of development it would be replaced within three to four years. The Industrial Revolution really began in 1945.

The average industrial growth rate throughout the world between 1963 and 1968 was 7 per cent per year. This gives an overall doubling time of fourteen years. Unfortunately, of course, this growth is not distributed evenly. We are in a situation in which the rich get richer and the poor get children! If we compare the growth rate of populations with the growth rate of the gross national product (GNP) in particular countries we can see quite clearly the way in which what should represent increased material prosperity is being swallowed up:

Average growth rates 1961–8

Percentage	POPULATION	GNP
China (Mainland)	1.5	0.3
India	2.5	1.0
USSR	1.3	5.8
USA	1.4	3.4
Pakistan	2.6	3.1
Indonesia	2.4	0.8
Japan	1.0	9.9
Brazil	3.0	1.6

All of this industrial growth is dependent on resources of minerals and fuels, most of which are non-renewable. That is to say, once they are removed from the ground and processed, they are gone for ever. Some minerals, particularly metals, can be used and re-used and no doubt recycling will become increasingly important in the future. However, as the expense and complication of extraction or recycling increases, so does the amount of power consumed, and at the present time most of this power is derived from fuels that were fossilized millions of years ago. They are not being replaced. Once burned they are lost. The other mineral fuel source, uranium, is also in limited supply. Preston Cloud, professor of biogeology at the University of California at Santa Barbara, has predicted* the exhaustion of the entire world stocks of crude oil, natural gas and uranium by the year 2000.

How is such a calculation made? The world's stocks of most minerals is known with a fair degree of accuracy. Thus we can begin the calculation knowing how much of a resource we have. We know, too, the rate at which it is being consumed. By relating the rate of consumption to the quantity in reserve we can calculate the number of years the reserve will last. This figure is known as the static reserve index (SRI). The SRI is the figure often quoted for the lifetime of resources. However, the rate of consumption is not a constant. If it is related to industrial and population growth, together with the trend it has shown over a number of years, and this new value is compared with the known reserve, the resulting lifetime for the

* Mined Out!, by Preston Cloud. *The Ecologist*, 1.2, August 1970.

World Energy consumption in 1980

Developed countries	A^a	SOLID 10^{12} $kWh(t)^c$	SOLID *percentage of World consumption*	LIQUID 10^{12} $kWh(t)^c$	LIQUID *percentage of World consumption*
USA	3.5	5.0	17.3	9.4	25.3
Canada	5.5	0.3	0.9	1.2	3.2
W. Europe	4.0	2.7	9.4	9.2	24.9
E. Europe	4.6	3.6	12.5	1.5	3.9
USSR	6.5	5.7	19.7	5.2	14.0
Japan	7.9	0.5	1.9	3.5	9.5
Oceania	4.8	0.4	1.3	0.4	1.2
TOTAL	4.7	18.2	63.0	30.4	82.0
Developing countries					
Comm. Asia	7.6	7.3	15.4	0.7	2.0
Other Asia (except Japan)	8.5	2.3	3.1	2.4	6.5
Africa	6.5	0.9	0.6	0.7	1.9
Other America	7.4	0.2	7.9	2.8	7.6
TOTAL	7.7	10.7	37.0	6.6	18.0
World TOTAL	5.2	28.9	100.0	37.0	100.0

Developed countries	GAS		HYDRO[b]	
USA	8.3	41.9	0.34	18.1
Canada	1.0	4.8	0.22	11.6
W. Europe	2.1	10.3	0.46	24.1
E. Europe	0.7	3.5	0.02	1.2
USSR	5.9	29.8	0.29	15.3
Japan	0.1	0.4	0.11	5.7
Oceania	0.1	0.7	0.04	2.1
TOTAL	18.2	91.4	1.48	78.1

continued

World Energy consumption in 1980

	GAS		HYDRO[b]	
Developing countries	10^{12} *kWh(t)*[c]	*percentage of World consumption*	10^{12} *kWh(t)*[c]	*percentage of World consumption*
Comm. Asia	—	—[d]	0.04	2.2
Other Asia (except Japan)	0.4	2.2	0.14	7.3
Africa	0.2	0.9	0.06	3.0
Other America	1.1	5.5	0.18	9.4
TOTAL	1.7	8.6	0.42	21.9
World TOTAL	19.9	100.0	1.90	100.0
Developed countries	NUCLEAR[b]		OVERALL	
USA	0.98	52.0	24.0	26.8
Canada	0.05	2.8	2.8	3.0
W. Europe	0.63	33.5	15.1	16.8
E. Europe	0.04	2.1	5.9	6.6
USSR	0.04	2.1	17.1	19.1
Japan	0.11	5.6	4.3	4.9
Oceania	0.01	0.3	1.0	1.1
TOTAL	1.86	98.4	70.2	78.3

continued

[a] Col. A contains the projected average annual percentage of growth in energy consumption for 1965–80.
[b] Hydro and nuclear electricity are equated, converted from metric tons of coal to kWh at the UN rate of 1,000 kWh per 0.125 mtce.
[c] Converted from metric tons coal equivalent by using 27.3×10^6 Btu/mtec. and 0.293×10^{-3} kWh(t)/Btu
[d] Unknown, but believed to be small.

Developing countries	NUCLEAR 10^{12} $kWh(t)^c$	NUCLEAR percentage of World consumption	OVERALL 10^{12} $kWh(t)^c$	OVERALL percentage of World consumption
Comm. Asia	0.01	0.4	8.1	9.1
Other Asia (except Japan)	0.01	1.0	5.2	5.8
Africa	—	—	1.8	2.0
Other America	0.00	0.2	4.3	4.8
TOTAL	0.03	1.6	19.4	21.7
World TOTAL	1.89	100.0	89.6	100.0

(Harmstadter, J. *Energy and the World Economy*)

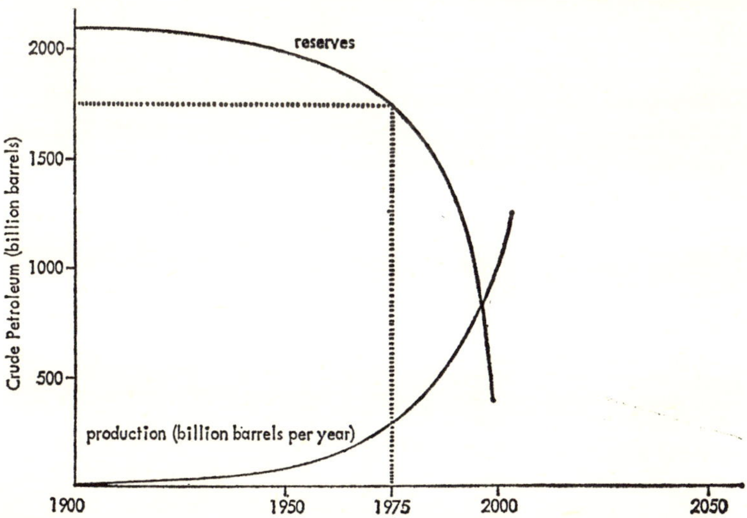

Reserves of crude petroleum. It is a feature of exponential curves that they show little change for a prolonged period, followed by a very rapid rise or fall. In this graph, which plots consumption against known reserves it is possible to see another feature of exponential curves: in 1975, with only fifteen years before demand exceeds supply, only 12½ per cent of the world's reserves have been consumed.

resource is known as the Exponential Reserve Index (ERI). An exponential growth in demand will reduce the lifetime of a reserve by a very large factor. If, for example, we take a resource with an SRI of 10,000 years and then assume an exponential increase in the rate of consumption of 2 per cent per year, the ERI is 250 years! If, at some stage before the stocks are exhausted completely, the reserve that remains is doubled – let us assume that new sources are discovered, or there is a major breakthrough in extraction technology which permits us to mine reserves that were previously uneconomic – the effect is to add seventy-five years to the ERI. In other words, doubling the stocks may add no more than seventy-five years to the time taken to use up the resource completely. The cost of extracting the resource can also be estimated and it is found, not surprisingly, that as the reserves are reduced – the most easily accessible being used first – the cost rises exponentially.

SRIs and ERIs have been compared for a number of resources. The results are disturbing:

	SRI years	ERI years
Aluminium	175	34
Chromium	560	92
Coal	900	595
Cobalt	155	45
Copper	40	25
Gold	17	14
Iron	400	73
Lead	15	12
Manganese	160	49
Mercury	13	13
Molybdenum	100	27
Natural gas	35	14
Nickel	140	30
Petroleum	70	20
Platinum	20	12
Silver	20	17
Tin	25	17
Tungsten	40	36
Zinc	18	16

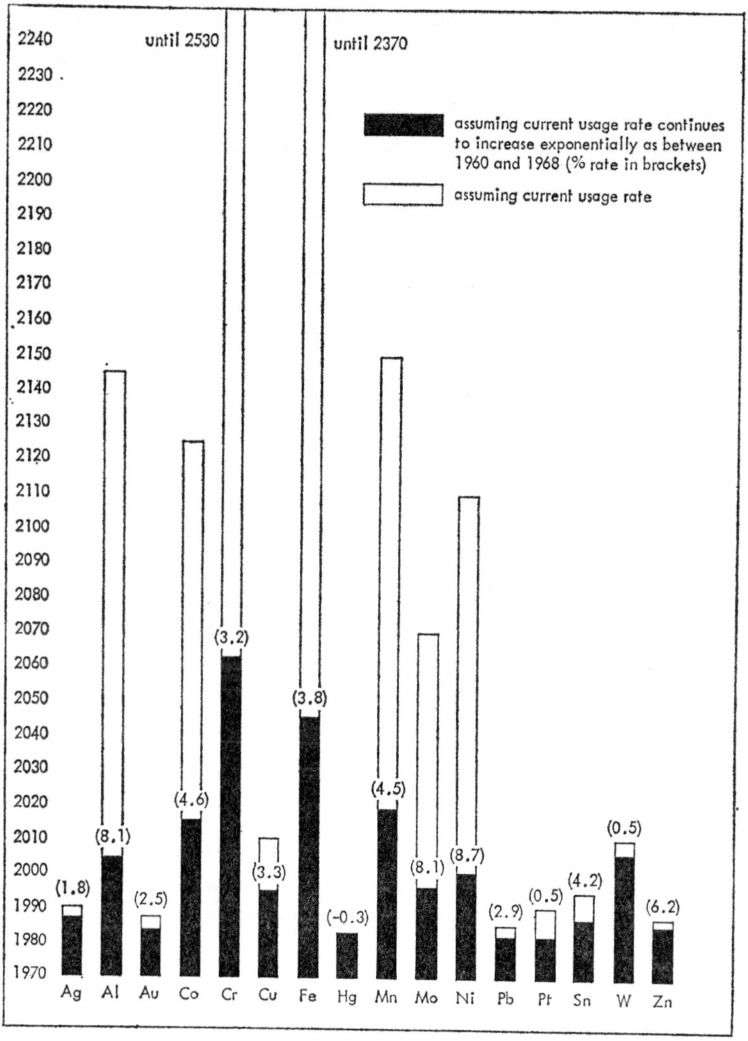

Static and exponential reserves of metals.

The implication is that continued industrial expansion may bring a certain amount of short-term material prosperity to a few people in some of the richer countries, but that in the long term it can serve only to bring nearer the date by which the basic materials on which it is based are exhausted. Since it is the richer countries that are most heavily dependent on mineral resources, it is they that will suffer first and most severely.

Relating this to population growth, it is clear that within the foreseeable future – probably within the next fifteen to twenty years – not only may there be a shortage of food that will make it more difficult and expensive for the industrialized food-importing countries to buy food abroad, but it is certain that they will find the greatest difficulty in continuing to manufacture the goods they would use to pay for such food. As the price of commodities rises in response to a diminution in supply, major world-wide inflation may cause an economic collapse that will accompany the final collapse of industry as we know it. It becomes highly relevant, therefore, to consider the size of population that can be fed within the sustainable food production capability of the area in which it lives. The days of substituting the production of machinery for that of food and of living by trading are coming to an end. Unless governments accept this and introduce policies to regulate populations at appropriate levels, severe food shortages may be unavoidable, at least in some areas. There are industrial countries that could feed themselves, but they are few and they certainly do not include Britain. By the end of the century, food availability in this country may have been halved and the historical evidence suggests that the effect of this would not be to halve the food intake of each of us, but to widen the gap between rich and poor until large numbers were reduced to subsistence level and less.

The disease is growth. Populations throughout the world and industrialization in parts of the world are growing at an ever-increasing rate. It is a certain recipe for disaster. Such growth cannot be sustained for more than a few years and the present material affluence of the developed countries, and their talk of annual economic growth rates of 3, 4 or 5 per cent may, in time, be seen as no more than a phenomenon of the '60s and early '70s, the years of profligacy when, with insane abandon, we devoured the planet and, with it, the future.

The situation is not hopeless, of course. An increasing number of influential citizens in all countries is aware of the impending crisis and all over the world groups of people are meeting to prepare strategies for changing the direction of our civilization. If they are successful, if governments and the majority of the peoples of the world can come to recognize the need for radical changes, then the future could offer many of the benefits of modern technology and enough food for everyone. Material standards of living in the countries that are now industrialized would be rather lower, but throughout most of the world they would be a good deal higher. New values and goals would be developed that might provide far more satisfaction in human terms. One way or another, the materialism for which our civilization is criticized so often is coming to an end. We are moving rapidly into a post-industrial era in which human, cultural, spiritual values may replace those of commerce, and the question asked of a new venture may be 'What will it contribute to the quality of life of the community as a whole?' rather than 'Will it be profitable?' It could happen. We possess the knowledge to make it happen.

Then again, there is an alternative. Human history might come to an abrupt end early in the next century.

3

The 'Green Revolution'

The world population is increasing at a rate of 1.9 per cent per year, but this average is not distributed evenly. In 1965, the developed countries of Europe, the USSR, North America and Japan had rather more than 1,000 million inhabitants. A further 800 million lived in the communist countries of Asia. Some 1,500 million lived in the developing countries. The world average of 1.9 per cent per year includes figures that are much higher. The average growth in the years 1952–4 to 1967–9 was 2.52 per cent in Latin America, the Far East, the Near East and Africa. Growth exceeds 3 per cent in all the countries of Central America, four countries in South America, five countries in the Far East, two countries in the Near East and six countries in Africa. The effect of these disparities will be an increase in the populations of the developed countries of 25 per cent in the twenty years between 1965 and 1985, and an increase of 60 per cent in the developing countries. Or, to put it another way, of every 100 additional people born in the world between 1965 and 1985, 85 will live in developing countries.

If the diet of the peoples of the developing countries is to be maintained at its present level of adequacy, then obviously food availability must increase at least as fast as population. Between 1956 and 1966, food production in these countries increased by 2.7 per cent a year. This represents no mean achievement. If we take the years 1952–6 as a base and allot food production during that period a value of 100, total agricultural production up to 1969 has increased by between 40 and 60 per cent:

	1956	*1960*	*1965*	*1969*
Latin America	109	118	140	151
Far East	107	121	133	154
Near East	110	124	145	163
Africa	106	121	135	143

The figures look impressive, and they suggest that people should be eating as well as they were, even if they are not eating much better. Unfortunately, food production is not the same thing as food availability. We must take account of wastage after harvest and losses due to inadequate and inefficient distribution. When we do, we find that the situation generally is deteriorating. In many cases people actually get less to eat than they did in the 1930s! It is difficult to be precise, but probably between half and two-thirds of the people in the developing countries are nourished inadequately at the present time.

	CALORIES	PROTEIN
	available as percentage of requirement	
North America	114	123
W. Europe	111	116
Japan	94	108
Oceania	126	127
S. Africa	117	124
USSR	116	figures not available
E. Europe	118	–
Latin America	109	105
Central America	106	101
South America	111	106
Central Africa	91	68
East Africa	93	99
West Africa	97	93
Near East	92	105
N. Africa	88	84
India	90	86
Pakistan	85	89
Other Asian countries	96	91

In this table, 100 represents bare subsistence in terms of populations as a whole. In fact, the tendency is for the wealthier sections of communities to eat well, so depriving the poor. Thus a food availability index near or below subsistence level indicates a small, well-fed population and a large population fed less well even than the figures suggest.

It is not enough, even then, for food availability to increase at the same rate as population growth. It must increase faster, and for a very important reason. If farmers are to grow more food, they must be able to sell it to those who need it. Those who need it must have the money to buy it. Therefore, before there can be any permanent increase in agricultural production there must be an improvement in the general level of prosperity, without which the additional food will not reach those in greatest need. It is an ironical fact that one of the necessary stages in providing more food for the hungry is to create a market for food! Without this existing markets will be glutted; those who can afford to buy enough presumably do so already and they are unlikely to buy much more. So agricultural development can be checked at the start unless it takes place in a favourable economic climate. As soon as earnings begin to rise and the economy as a whole begins to improve people will be able to buy more than they need for mere survival. There will be an increase in expectation. While there is a need to create a demand, the potential of this demand once it has been generated is larger than could be supplied by increases in production up to a level of dietary adequacy.

If food availability were to match population growth only there would have to be an increase of about 80 per cent in production between 1965 and 1985. If there is an increase in incomes leading to a greater demand, this required increase rises to 140 per cent. As we have seen, neither of the two targets is being met at the present time and more recently it has begun to appear that they themselves are low and that an increase in production of 140 per cent by 1985 would do no more than match population growth.

According to figures prepared by the FAO,* a population growth of 2.6 per cent a year will call for a growth in food production of 3.9 per cent per year. It is the difference between the actual growth rate of 2.7 per cent and the required growth rate of 3.9 per cent that represents the world food problem. There is a shortfall of 1.2 per cent per year. Just as the growth rates are exponential, so is the shortfall, and over a number of years it represents a great deal of

* *The Provisional Indicative World Plan for Agricultural Development*, FAO, Rome, 1969.

food, a great deal of hunger. According to a statement issued in Rome by the FAO on the 12 February 1971:

'More than 12 million children under the age of five in developing countries are in danger of death according to a report of the Protein Advisory Group which met here this week at the headquarters of the Food and Agriculture Organization.

'The Group, which is sponsored by FAO, the World Health Organization and the UN Children's Fund reviewed the current protein situation in the world under the chairmanship of Prof. Nevin S. Scrimshaw (USA).

' "Twenty per cent of all children between birth and five years of age in developing countries show definite signs of protein malnutrition and three per cent are in danger of death, if they are not given an adequate diet and proper medical treatment," the report says. There are more than 420 million children under the age of five in the developing countries.

'In a statement, the Group urged "recognition of the fact that children experiencing severe malnutrition in infancy show not only permanently reduced physical growth and development but also impaired learning and behaviour."

'Even where statistics show adequate protein resources on a per capita basis, economic inequities leave large masses of the populations of many developing countries with dangerously low protein components in their diet.

'In 1963, 43 countries with a population of 900 million had protein supplies below the recommended level, the Group reported, but since then the trend has not been a reassuring one. Although the protein ration in developed countries exceeds requirements by some 30 per cent and a continuing increase has brought the per capita ration in Latin America to 105 per cent of demand, the situation in Africa is disquieting. The current protein ration only meets 90 per cent of average requirements. In the Near East, the ration, formerly adequate, at least on a per capita basis, is showing a tendency to decline.

'Even if production targets to bring protein supplies up to 10 or 15 per cent above theoretical per capita needs were met, the Group warns that the vulnerable population groups – the poor, women and children – probably still would suffer from malnutrition because of mal-distribution and cultural factors.'

A further statement issued on 1 March announced that Pope Paul would bless more than 100,000 people expected to take part in a 'Walk for Development' in Rome on 9 May 'and, through them, the millions of others walking in similar marches on five continents on that day.'

The body constituted to find solutions to the problem – if solutions there are – is the Food and Agriculture Organization, the largest of the United Nations' agencies. Like the others, it works under severe political restraints. It is wholly dependent for funds and for access to the territories within which its work is undertaken on the goodwill of governments. If the problem is being exacerbated by the policies of governments, as is frequently the case, the FAO finds it very difficult to criticize. Some years ago, Lester Pearson, a former prime minister of Canada and an ardent and active worker for the FAO, suggested that each of the developed countries should allot one per cent of their GNP to aid developing countries. The aim of this aid would be to help these countries to feed themselves. None of the developed countries has responded fully and when economic crises threaten at home it is frequently foreign aid that is the first casualty as the rich governments reduce their spending. To be brutal, there are no votes in aid.

In 1963 the FAO held a 'World Food Congress' in Washington, DC. As a result of this Congress, it was decided to prepare a plan for agricultural development throughout the developing world for the period up to 1985. The plan was published in 1969 with the title 'Provisional Indicative World Plan for Agricultural Development,' or 'IWP' for short. It was this plan and its implementation that was nicknamed 'The Green Revolution.' The name has stuck, but although the FAO uses it itself, its Director-General, Mr. A. H. Boerma, has said that he did not invent the name, he does not like it, and he would like to see it dropped.

It is easy to see why. It is too facile. It makes it too easy for people, particularly people in the rich countries, to believe that the problem is solved, that there is no more need for their concern. It sounds simple and dramatic. In fact it is neither. The IWP is a long document. It is subtle and complex. Unfortunately, for reasons which we will consider in more detail later, it is very unlikely that it will work, and if it should be successful it might well create insuperable ecological problems, leaving those it aims to assist worse off than ever.

What is the plan? The developing countries need more food. If they were to import it, the cost would reach 26,000 million US dollars by 1985, provided prices remained at their 1962 levels (providing, too, that the food were available for purchase in such quantities.) In 1962, the countries of the third world found great difficulty in meeting a food import bill of 3,000 million dollars. We can discount the importation of food as a permanent solution.

Nevertheless, there are countries that have attempted to solve their problems in this way. Frustrated at the slow development of agriculture, they have tried to short-cut the development process by attempting a rapid industrialization that they hoped would lead to economic 'take off.' In a very few areas, rich in mineral or oil deposits, this has worked. Elsewhere it has proved disastrous. Factory industry is seldom labour-intensive and it provides relatively few jobs. If investment is diverted from agriculture to pay for it, agricultural development is slowed. This leads to a rise in rural unemployment. This reduces demand for the products of agriculture and of industry. People drain from the countryside into urban areas in the hope of finding work, but the work does not exist and the cities cannot accommodate them. Food supplies into the towns decrease as urban food demand increases. The flow of materials through industry flags, industrial development slows down and exports decline while food imports rise. The trade balance swings heavily into the red and a vicious spiral is created. The situation grows worse and worse. It is interesting to remember that the industrial development of all the countries that are rich today began with a secure and prosperous agriculture. It was generally the surplus wealth created by farmers and landowners that provided the initial capital for industry. Historically, it seems that a sound agriculture is vital to all subsequent economic development.

Agriculture must provide all the food that will be needed in the immediate future and it must also earn and save foreign exchange, and provide additional employment as the population increases as well as further employment in the 'agro-allied' industries.

If we are to increase agricultural production, we must begin with cereals. It is cereals that provide the staple foods and the feedstuffs for livestock. Moreover, they are psychologically important. 'Bread' is often synonymous with 'food.' Even we talk of bread as the 'staff of life.' The IWP therefore devotes a great deal of attention to ways in which cereal production may be increased.

There are two possible lines of approach. Existing farmland may be cultivated more intensively, or new 'virgin' lands may be brought into production. The IWP opts unequivocally for the former. To bring new land under the plough is expensive and, initially at any rate, the returns are low. The capital could be spent more usefully in improving methods on existing farms.

There is another reason. The amount of available land is limited. New lands have been brought into cultivation throughout history. In 1882 the remaining tillable land in the world was estimated to amount to 4,700 million acres. By 1952 this had shrunk to 950 million acres. Furthermore, it is usually the best land that is used first. As farming spreads into the virgin areas it is spreading into inferior land. Thus, in 1882, 85 per cent of the world's farm land was classified as good; by 1952 the figure had fallen to 41.2 per cent. This extension into poorer land has caused, or accelerated, soil erosion. This may lead eventually to the land being abandoned. In the long run the total area under cultivation remains much the same. According to Georg Borgstrom, professor of food science and geography at Michigan State University*:

'We thus find that about 40 per cent, i.e. two-fifths, of the land surface, either consists of deserts and wasteland or has been taken for rather permanent human habitation. Furthermore we find that 20 per cent of the land surface is covered by inaccessible forest almost entirely located in areas and terrain which is difficult to reach for forestry operations as hitherto known . . .'

Borgstrom's calculations indicate that of the whole surface of the globe, only 17.4 per cent is suitable for food production.

'It is therefore a safe prediction that for a population twice as large as at present, which experts agree we will have prior to the year 2000, the output will not fill minimum vital needs.'

* *Too Many; a study of Earth's biological limitations,* by Georg Borgstrom, Macmillan, 1969.

Land can be treated as any other resource is treated: by comparing the present allocation of cultivable land per capita throughout the world with population growth it is clear that soon there will be no new land left. The IWP itself confirms this:

'In Southern Asia ... in some countries in Eastern Asia, in the Near East and North Africa, and in certain parts of Latin America and Africa ... there is almost no hope for expanding the arable area. ... In the drier regions it will even be necessary to return to permanent pasture the land which is marginal or sub-marginal for cultivation. In most of Latin America and Africa south of the Sahara there are still considerable possibilties for expanding the cultivated area, but the costs of development are high and it will be often more economical to intensify utilization of the areas already settled.'

In 1965 neither the USSR or China had any remaining agricultural reserve apart from forest or pasture. Of their total land areas, Europe still had 1.4 per cent that was potentially tillable, North and Central America 4.7 per cent, South America 2.9 per cent, Asia (apart from China) 5.2 per cent, Africa 3.7 per cent and Oceania 0.1 per cent. We can forget about expanding into new lands.

In many rural areas throughout the world there is already considerable unemployment. More people live on the land than can usefully be employed on it, but they are unable to leave because there is nowhere for them to go and employment prospects for them are no better anywhere else. The intensification of production on existing farmlands would bring more employment.

The IWP aims, therefore, to begin with an increase in cereal production on existing farms. The most important dietary deficiency arises from a shortage of protein, as we have seen, and in most developing countries the major protein source is vegetable. Calorie requirements vary somewhat according to climate and cultural preferences and are known with some accuracy. Less is known about protein requirement, either in quantity or in quality, but the FAO estimates a daily protein requirement of 48.8 grams per person. In 1963, the average daily per capita protein consumption in India was 50.1 grams of which 6.4 were of animal origin. In Mauritius, of a total of 49.1 grams, 13.4 were of animal origin. By 1968 the figures for India had fallen to 47.8 and 6.1 grams, and in Mauritius they had fallen to 46.2 and 12.5. These changes were general.

If the protein content of cereal crops were to be increased, this would produce a very profound effect on the quality of diets throughout the areas affected. A few more grams of protein in their rice could improve at one stroke the protein/calorie balance of the diet of many Asians and might eliminate the protein deficiency in Asia. According to the IWP:

'If half the rice consumed in Asia were replaced by 1985 by a new variety with a protein content of 11 per cent, the entire region's protein supply would shoot up by 2 million tons and average individual consumption of protein would rise from 60 to 65 grams a day.'

The plan calls for the development of new, high-protein varieties. This will not only increase protein available directly. It will also improve the quality of animal feedingstuffs and so increase the proportion of meat protein in the diet.

Quality classification of Tilled land

Percentage	1882	1952
Good	85.0	41.2
Half of original humus lost	9.9	38.5
Marginal soils	5.1	20.3

(Borgstrom, G., *Too Many*)

Pesticides needed to increase Food Production on acreage now under cultivation in Asia except Mainland China and Japan), Africa, and Latin (America by percentages of increase indicated

Percentage increase in agricultural production	Tonnage needed (metric tons)
—	120,000
10	150,000
20	195,000
30	240,000
50	342,000

60	412,000
70	475,000
80	558,000
90	640,000
100	720,000

(President's Science Advisory Committee, 1967)

World average rates of increase for the period 1951–66 for selected aspects of Human activity related to Food production

	Percentage
Food	34
Tractors	63
Phosphates	75
Nitrates	146
Pesticides	300

Rates in constant dollars (UN Statistical Yearbook 1967)

Changes in Land Utilization 1882–1952
(*in thousand million hectares*)

					CHANGE	
	1882	percent-age	1952	percent-age	1882-1952	percent-age
Forest	5.2	45.4	3.3	29.6	−1.9	− 36.8
Desert and wasteland	1.1	9.4	2.6	23.3	+1.5	+140.6
Built-on land	0.87	7.7	1.6	14.6	+0.73	+ 85.8
Pastures	1.5	13.4	2.2	19.5	+0.7	+ 9 41.
Tilled land	0.86	7.6	1.1	9.2	+0.24	+ 24.5
	9.53	83.5	10.8	96.2	+1.27	+ 12.8

Area not especially utilized	1.81	16.5	0.27	3.8	−1.54	− 79.9
TOTAL	11.34	100	11.07	100	−0.27	− 2.4

(Doane, R. R., 1957. *World Balance Sheet*)

To a very large extent the 'Green Revolution' is based on a series of new cereal varieties that have been developed and that the FAO aims to introduce throughout the third world. It was Jonathan Swift who said that 'who ever could make two ears of corn or two blades of grass to grow on a spot of ground where only one grew before would deserve better of mankind than the whole race of politicians put together.' The work on the development of the new varieties began in Mexico as a result of a project carried out by the Rockefeller Foundation in collaboration with the Mexican Government. The leader of the research team was a fifty-seven-year-old Norwegian-American from Iowa, Dr. Norman Borlaug. He was awarded the 1970 Nobel Peace Prize. The Nobel Committee said of him:

'As a plant improver, Dr. Borlaug has created a technical breakthrough in wheat production that makes it possible to remove hunger in the developing nations in the course of a few years.'

The eulogies were understandable. This was to be the 'Green Revolution.' The problem was all but solved. No wonder the world was enthusiastic. Dr. Borlaug's honour was richly deserved. Only there were a couple of snags:

What Dr. Borlaug's team had done was to develop strains of cereals that will grow well in the tropics and which are highly responsive to inputs of fertilizer and water. Basically the solution is based on increasing input in order to increase output, a sound enough approach one might think. The difficulty arose because the traditional varieties, which are well-adapted to life in their own environment, are long-stemmed. When fertilizer was applied to them the stem was unable to support the weight. The plants responded to the fertilizer by growing too tall. The ear grew more heavy and the plant toppled over. Sometimes this is sufficient to bring the seed into contact with the ground. This phenomenon is called 'lodging.' It is a problem well-known to all farmers. It may mean the crop cannot be harvested. If left, or ploughed in, there is a chance that some of the seeds might germinate, so spoiling the following crop.

The new varieties are short-stemmed, and as new varieties appear each seems to be shorter than the last. The first wheats were 36 to 42 ins high, the second 30-38 and the third only 26 to 30 ins. They have given yields up to 20 per cent higher than traditional varieties given the same treatment. Indeed, they seemed so promising that European farmers began to wonder when they would be able to obtain seeds. Unfortunately, the answer is 'never.' These plants would not grow well enough in temperate conditions. The development was taken up by rice breeders working in the Philippines. They, too, developed a whole new series of rice varieties. One of them, IR8, became known as 'miracle rice.' In the Quezon Province of the Philippines it yielded 10,000lb per acre, compared with the 1,330lb produced by the traditional varieties. End of the problem? Not quite!

The new varieties require inputs, first of water and fertilizer. To provide more water we will need improved irrigation – although even then there are disadvantages. This costs money and the developing countries do not have the money. It must come from the rich countries. So far it has not.

What is the aim of irrigation? It is to redistribute such fresh water as exists. It will not produce water. In many of the areas where agricultural improvements are planned, there is already a shortage of water. If more is diverted to agriculture, either it must be taken from other users, or it must be derived from wells sunk into the ground water, which would deplete the natural stocks. If we take water from other users there will be further shortages. If domestic users give up some of their water, shall it be drinking or washing water? Are we suggesting that the satisfying of people's hunger will add to their thirst? It is not an easy choice. It may be one between a reduction in standards of hygiene, which could increase the incidence of disease, or thirst. People might prefer to be hungry rather than thirsty and they might not part with their water. Perhaps we could take the water away from industry? In the industrial countries, industry is by far the largest user of water. This is less true in developing countries and there some industrial development is needed if people are to be prosperous enough to buy the additional food. Shall we produce more food, but slow down the development of the means to buy it?

Perhaps we should tap the reserves further. This, too, will cause problems – less immediate perhaps, and less dramatic at first, but more serious in the long run than either of the two previous alternatives. We will consider the availability of fresh water in a later chapter. For the moment it is enough to point out that if water is taken from underground reserves and dispersed on to the surface, some of it will evaporate. Obviously, the amount lost in this way will vary according to the air temperature and humidity. It is reasonable to assume that the water would not be brought up at all unless the weather was dry and sunny: the very worst time for evaporation. The reserves will diminish faster than they can be replaced, so rendering the 'mining' of water no more than a temporary phase.

If the additional water is taken from rivers, then their level will fall and pollution will increase. This is because pollutants discharged into a river are diluted by the river water. If the volume of water in the river is reduced, the concentration of pollutants will increase and so the river will be more severely polluted. This is a world-wide problem, because throughout the world we use rivers as sewers. As the degree of pollution increases, the water will become less suitable for domestic use. If water is to be taken from rivers, the further downstream the better. The damming of rivers upstream will supply large reservoirs that can be used to dampen down the effects of seasonal variations in rainfall and they can provide a source of hydro-electric power, but they produce problems, too, which we will consider in more detail later. It seems that at best large dams are a short term expedient.

All of this leaves us a little dry. Is the water available to grow the new varieties? In the long term it is doubtful. In the short term the capital has not been made available to provide the irrigation that might lead to a quick, even if short-lived, success.

The high-yielding varieties are defined as those that give a linear response to fertilizer applications up to 90lb per acre. Indeed, it is misleading to talk of them as 'high-yielding' varieties. The IWP is at pains to make this clear. They are highly responsive to applications of fertilizer and water, and this is not the same thing. Without fertilizer they do not give high yields.

Ninety pounds per acre is a low rate of fertilizer use by European or North American standards. In South Eastern England, for example, farmers use more than 200lb of artificial fertilizer on each acre of arable land. However, the new varieties are also quick-growing. The 10,000lb of IR8 was obtained by growing several crops in the same season – multiple cropping. In some places the new cereals will give as many as three crops in one year. To do this, the 90lb per acre of fertilizer that they require must be multiplied by three. Two-hundred-and-seventy pounds per acre is a larger requirement than that of some of the most intensive arable farmers in the world!

Arable farmers in the developing countries in 1968–9 consumed an average of 11½lb of fertilizer per acre. The IWP is asking some of them to increase this consumption by nearly twenty-four times. To do so will require heavy capitalization. The capital can come only from the rich countries. It has not been given.

We will consider each of the major inputs in more detail later, but not only is the necessary fertilizer not being made available at the present time, it is unlikely that it could be. It is far from certain that even if the world could produce fertilizer in these quantities, it could sustain such a level of production for more than a few years.

So the fertilizer is not being made available and it seems unlikely that it will, or can be. If it were it might well lead to further long-term ecological complications that would leave the intended beneficiaries worse off than they are now.

The introduction on a large scale of any new plant variety will cause gross changes in the ecology of the area, which may result in an increase in crop pests, weeds and disease. If we wish to intensify agriculture in this way this is unavoidable and we must accept it. It seems, however, that the new hybrids are more than usually subject to pests and diseases. This may be due in part to the relatively high density at which they are grown, which will offer greater opportunities to pests and disease organisms, and it may also be due to the high degree of specialization with which they have been bred. There is always a danger that by breeding for one particular quality others are sacrificed. We have only to look at pet dogs to see that by breeding for an exaggerated appearance we sometimes produce a weak animal that would find it difficult or impossible to survive in the wild. The same is true of plants. By selecting for short stems and rapid growth the breeders may have increased the susceptibility of the plants to pest and disease attacks.

Whatever the reason, the crops will not thrive without an increased input of crop protection chemicals – insecticides, herbicides, fungicides, and so on. In the period 1962–4, 20 per cent of the world's pesticide output was used in the developing countries, on 70 per cent of the world's farm land. If pesticide consumption in the third world is to increase, capital will have to be made available for their purchase. It is unlikely that many developing countries will be able to manufacture for themselves the range of materials in use at present. The capital has not been forthcoming.

If pesticides were to be introduced on the scale required, there is excellent reason to suppose that they would create problems far worse than those they would solve, and there is some doubt as to whether they could solve the problems even then for more than a short period. There is further reason to suppose that not only does the world not produce sufficient for everyone, but that it cannot.

The final requirement is, perhaps, the one that can be supplied. The new varieties require, in the words of the IWP: 'a continued flow of suitable new varieties, resistant to the major pests and diseases, and capable of high yields in response to the application of modern farming technology. This implies both a multi-disciplinary breeding and research programme to produce and test the varieties; and a well-organized multiplication and distribution programme to ensure that quality seed is available to the farmers in adequate quantity.'

In other words, a constant supply of seed.

The problem is more than an agricultural one and its solution calls for the efforts of more than agriculturists. If the 'Green Revolution' were successful and yields were increased, we must be sure that the benefits reach those for whom they are intended. The aim is to feed the hungry, not to make the well-fed still fatter. This sounds obvious, but the social and economic structure within recipient countries may not be adequate to ensure the equitable and efficient distribution either of the inputs or the outputs. There was a riot in India, which resulted in the death of a number of agricultural

Pesticides

economic structure

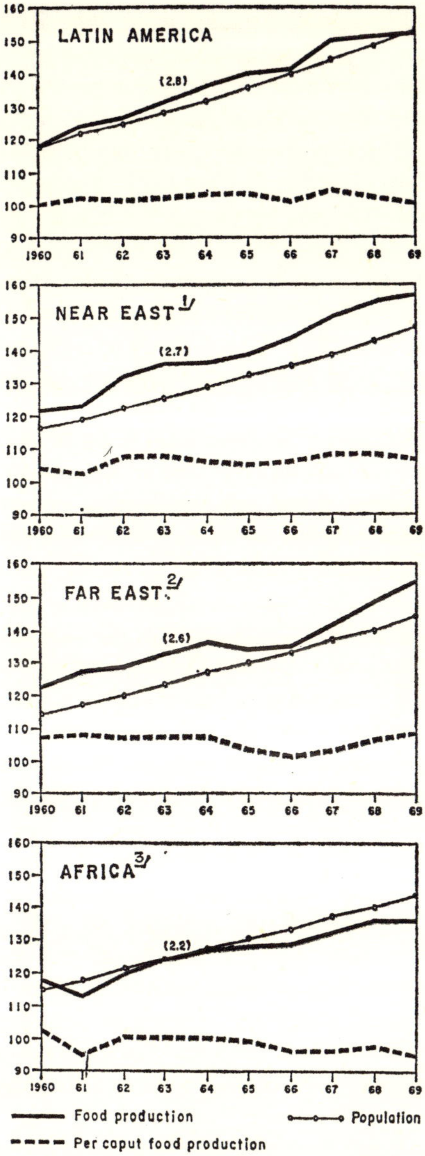

Increases in food production and population growth in developing countries. Figures in brackets show annual rates of increase in food production 1956–8 to 1966–8.

workers, caused by the introduction of blackleg labour by a farmer unwilling to pay higher wages for the harvesting of an increased yield. He refused to share his additional profit with his workers. There have been cases of tenant farmers who introduced the new varieties, grew larger yields, and found that their landlords had increased their rents.

We have discussed some of the limitations that reduce the entire 'Green Revolution' concept to what sounds like a 'technological fix.' This term was coined by Nigel Calder to describe the quick, simple, neat, technological 'shot in the arm' that can produce an apparent immediate solution to a problem. It may be genuine. More likely it is based on an over-simplification of the problem, or even on the invention of a fictitious problem as a more amenable substitute for the real trouble. The impression given is that technology has cured yet another of our ills.

How is the 'Green Revolution' coming along? What is actually happening?

In spite of twenty-six years of effort by the FAO, people in the developing countries are now worse off than they were before the Second World War. Yet agricultural production has increased. In general, food production has more or less kept pace with population growth, but there are signs that the situation is deteriorating already. In 1969, for the first time in twelve years, there was no increase in the combined production of the world's farms, fisheries and forests. This was due largely to reduced output in the developed countries, where surpluses rather than shortages are the problem, but even so, with the exception of the Far East, there has been a fall in food output throughout the developing world. The FAO draws some comfort from the fact that the one region in which production did increase, the Far East, is the one where shortages had always been the most acute. Even there the increase was below the increase for 1968, suggesting that the rate of annual increase may be levelling off. The FAO figures exclude Mainland China, but there is a sign of hope in preliminary reports of a record 1970 rice crop in China. China is the world's third largest rice exporter and it is estimated that the 1970 harvest is 5 per cent larger than the 1969 harvest. According to the FAO's information, the increase is due to putting more land under cultivation, double cropping in some areas through the use of early maturing varieties of rice, better seeds and more advanced techniques. It does not follow necessarily that China will be able to increase its exports of rice in proportion to the increased harvest, but there is some ground for cautious optimism.

What about the new varieties? Where is the 'miracle rice'? Where they have been introduced, yields have increased. According to the FAO's annual report: *

'As against an overall increase of 4 per cent in the (Far East) region's total food production, the combined paddy crop of the region again expanded by 6 per cent, and that of wheat by 10 per cent. The total was held back mainly by smaller increases in the output of livestock products, and declines in the production of barley and pulses.'

The trouble is that they are not being introduced. In the Far East, the area where they have been planted most extensively, only 7 per cent of the total rice-growing land was sown to them in 1969. The FAO says that 'The rate and extent of their adoption have been influenced by a number of factors, including physical, ecological, institutional and structural constraints, economic pressures and consumer preferences.' In the Near East: 'A faster rate of expansion in the adoption of high-yielding varieties of cereals in the region remains unattainable until some of the major constraints are overcome. Aside from the limited areas under irrigation, and the competition for it by other crops, these include the shortage (except in the United Arab Republic) of well-trained and experienced wheat breeders who could plan and execute breeding programmes to meet future demands.'

It is water that is proving the first resource to be in short supply.

'In the longer run, the principle limit to the use of the existing high-yielding cereal varieties is the area under controlled irrigation, and the competition for it by other crops.'

Can this problem be overcome by developing varieties that grow well with less water? Biologically, this sounds like a contradiction in terms. If it were possible it would imply the reversal of the concept of increasing output by increasing input. Suddenly we would be seeking to increase output by an amount greater than that by which

* *The State of Food and Agriculture, 1970*. FAO, Rome, 1971

we were prepared to increase input – because the input is not available. This is to seek something for nothing. Like perpetual motion it is impossible.

Where the new varieties have been introduced further problems have arisen. The world food problem is first and foremost one of protein shortage. Traditional varieties of rice contain 7 to 9 per cent of protein. IR8 contains 5 to 7 per cent. Thus, the substitution of IR8 for a traditional rice represents a dietary deterioration. In any case, people prefer the traditional varieties. The new rices do not cook well. They have chalky spots and they become soggy. There are many instances of farmers growing the new varieties for a season or two and then reverting to those that they and their customers prefer. Meanwhile, those farmers who did not make the change benefit, because the introduction of the new varieties may force up the price of the preferred traditional ones.

The outlook is not promising. Populations continue to rise a little faster than food production and the 'Green Revolution' has run into serious difficulties before it can even start. Should it surmount these, it will encounter others. The chance of meeting the IWP target for 1985 is not good and, as we have seen, the target itself may have been set too low.

To a large extent the blame must lie with the governments of the rich countries. The FAO is well aware of the nature of the problems it faces. It knows, and has said, that they can be solved only if population growth ends and if large amounts of aid from the rich countries can be made available. Economic aid can come only from the rich countries, but population control can come only from them, too. As long as they demand that the developing countries limit their populations while they themselves refuse to accept that their own populations should be stabilized, the developing countries will reject their demands. To them it looks like genocide, an attempt to reduce the poor populations so that the rich may continue to prosper. They point out, with justification, that in terms of consumption of resources, the birth of a baby in Britain or America is a far more serious matter than the birth of a baby in India. The signatories to the Menton Declaration make the same point, but there is little evidence that governments are prepared to listen.

If the population of the world were to stop increasing now, could it be fed? Opinions vary and we will have cause to examine some of them when we consider the absolute limits to human expansion. According to some, there is enough food and the problems are of distribution. According to John McHale:* 'even if all the world's food production were shared out equally, there would still not be enough at the present time to feed everyone adequately, and this situation advances in seriousness day by day and year by year.'

There is a further danger that attempts to intensify production may accelerate soil erosion, so increasing the pressure on the better land until it erodes as well. This is one way in which the vicious spiral may be initiated that will take whole populations to the Malthusian limit.

There is real danger of major famines in the developing countries by the mid or late 1970s, which could increase the pressure on the industrial nations to become self-sufficient. In many cases this is impossible, so there may be famines in some developed countries by the turn of the century.

* *The Ecological Context*, John McHale, Studio Vista, 1971.

4

Water, water everywhere

Of course, there is plenty of water. 71 per cent of the surface of the planet is covered with it. We ourselves are made mainly from water. 70 per cent of our body weight is water. Those of us who live in Britain, where the average annual rainfall is around 35.5ins, frequently complain about the amount of water in our environment, although that does not deter us from travelling many miles to the coast to look at, swim in or sail upon still more. Surely, any suggestion that the world is short of water must be dismissed as no more than the ravings of the 'doom and gloom brigade'?

On Tuesday, 27 July 1971, an article was published in *The Financial Times* with the headline 'A race against the nation's thirst.' David Fishlock, the paper's science editor, was reporting on a one-day conference held in London at which Britain's water supply and demand were discussed. The summer of 1971 was dry and Sir William Goode, chairman of the Water Resources Board, told the conference that already it was too late to prevent a serious risk of water shortage within the next few years in a number of places. He mentioned Hull, Sheffield, the West Midlands, South Wales and Plymouth. He might have added parts of East Anglia, where periods of several weeks without rain are not uncommon and where already restrictions on the use of water during dry summer periods are commonplace.

We need a great deal of water. It is more vital to us than food. At a temperature of 76°F, an adult man requires a daily consumption of about 5 pints. Over a year this amounts to 5.5lb per day, or 2,000lb, of which 1,100lb is contained in his 'solid' food. But this is only the start.

We need water to produce food. Plants act as highly efficient pumping machines, taking water from the soil, incorporating some of it in their tissue and releasing the remainder into the atmosphere.

To collect the water available in the soil the plant has a complex root system which ends with tiny root hairs. A fully grown rye plant has a root system whose total surface area amounts to 2,420sq.ft and whose length has been measured as more than 400 miles. To this must be added 14 million or more root hairs whose total surface area may amount to more than 4,500sq.ft. Only a fraction of the water gathered by the roots is incorporated into the plant's tissue, so that the amount of water needed to produce, say, one pound of dry organic matter is truly immense. It varies, of course, from plant to plant, some requiring more than others. Rice consumes a great deal of water. It takes something like 2,200lb of water to produce one pound of dry rice plant. Wheat can grow well under relatively arid conditions. It takes only 350 to 500lb of water to produce a pound of wheat plant. I say 'rice plant' and 'wheat plant' deliberately, for these are the requirements of the plant itself. We eat only the seed of the plant and discard the rest, so in order to calculate the amount of water needed to produce a pound of rice or a pound of wheat grain we must allow for the fact that the bulk of the plant is discarded. In fact we need around 1,500lb of water to produce a pound of wheat grain for the miller, and some 10,000lb of water to produce a pound of rice ready for cooking – with water, of course.

These figures are dramatic, and they are meant to be. We take water for granted and overlook the vast amounts we consume. However, they are also slightly misleading. The portion of the crop plant that is discarded contains water, too, and some of this may be conserved. The crop residue may be ploughed into the top soil and the water re-released as the organic matter decomposes. Nevertheless, the fact remains that agriculture is a thirsty business.

In the industrial countries agriculture is a fairly minor water consumer. According to figures produced by the Water Resources Board, 25 per cent of all the water abstracted in Britain in 1969 went into the public water supply (2,900 million gallons per day, or mgd); 23 per cent was used by industry (2,710mgd); 0.5 per cent (60mgd) was used by agriculture; 0.3 per cent (30mgd) was used for other, miscellaneous purposes; and a giant 51 per cent (7,150mgd) was used by the Central Electricity Generating Board for cooling.

Again, some caution is necessary. Water tends to be used, rather than consumed. Water for cooling, for example, can be returned to the river from whence it came and provided the temperature is no higher than that of the river into which it is discharged, it has suffered little change. Water used by industry is also returned, but in this case it is usually polluted. It goes in as water and comes out as effluent with a load of chemicals. Losses do occur through evaporation and through the incorporation of water in manufactured products, but these have been estimated at no more than about 5 per cent of the total water used.

The fact remains that Britain may be about to face serious water shortages. At present we abstract some 13,000mgd and although much of this is returned relatively unchanged, the demand is increasing at about 5 per cent per year. It is our old friend exponential growth at it again. With our present population we consume about 37 gallons a day each. By the end of the century we could be consuming about 60, largely because of the increased demand from more advanced industrial processes that use more water and that frequently operate at higher temperatures, thus requiring more cooling. Whether this industrial development will take place is a matter for debate, but as projected, this will be the demand per head. Add to this the fact that by the end of the century our population will have increased to around 67 million and we see that within the next thirty years we will be using at least twice as much water as we use now. If we can get it.

Our annual rainfall of 35.5ins is ample to meet all these demands and more, but at present 15ins of it evaporates before it can be used and 18ins flow back into the sea. We must find ways of trapping and holding this water.

Almost weekly there is an outcry against plans to flood a valley to provide a reservoir. Land of great beauty and often of considerable agricultural value is sacrificed to meet the demand from the city and its factories. Estuarine barrages are suggested as an alternative to reservoirs. They may produce ecological changes in their own area, but at least they occupy no land. Meanwhile, the dry areas of the East and South-East of England derive much of their water from boreholes. This is equivalent to mining. It is the groundwater itself, the basic reserve, that is being tapped, and it is the diminution of this reserve as the level of the groundwater falls that leads to local shortages. There is nothing to fall back on. While the damp Englishman sits in his car waiting for the rain to stop so he can cross the beach to bathe, representatives of industrial, agricultural, town planning and conservation interests argue about ways to provide him with more water. We need water for everything. We use one or two tons of water to make a ton of bricks; 250 tons to make a ton of paper; 4 tons to make a ton of cement; 18 tons to make a ton of steel, 600 tons to make a ton of nitrate fertilizer.

Seventy-one per cent of the surface of the Earth is covered with water. The trouble is, of course, that most of it is salt water, and useless until it has been distilled, to remove the salt.

Water moves in a cycle – the hydrological cycle. Each year about 95,000 cubic miles of water evaporates from the surface of the sea and land. It moves with air masses and as it is cooled it forms clouds and eventually falls again as precipitation. About 71,000 cubic miles of its falls back into the sea. Of the 24,000 cubic miles that falls on land, 9,000 cubic miles returns to the sea through river systems and this balances the amount of water that the sea contributed to the land. The other 15,000 cubic miles evaporates from the land surface, either directly or after passing through vegetation. So the level of the seas remains constant. It is from this 9,000 cubic miles of fresh water that we must satisfy all our needs.

At the present time we cannot add significantly to this total volume of available fresh water. The best we can do is to attempt to increase the efficiency with which it is distributed. This is what we are doing when we plan to build reservoirs or when we undertake irrigation projects. We are not adding to water; merely re-distributing it.

The availability of water is a critical factor in crop production. Investigations in South Carolina have shown that in areas where the rainfall was an average of 2.82ins, soya bean production was between 23.7 and 26.3 bushels per acre. Where the rainfall averaged only 1.19ins, production was only 17.7 to 21.3 bushels per acre. In 1958 the United States winter wheat harvest was 48.5 per cent above the average and in 1962 spring wheat yielded 83.2 per cent more than usual. In both cases the reason was unusually high precipitation.

One of the most highly productive countries in the world in terms of yield of agricultural produce per acre of cultivated land is the

United Arab Republic (Egypt). For thousands of years the Nile was allowed to flood basins all over the delta and valley. When the floods subsided, openings were cut in the banks of the basins and the water flowed back to the river. The flood came in July, reached its height in September and then subsided quickly. The river brought silt washed down from its upper reaches and during the dry season the sun scorched the earth, killing off weeds. The fields were sown in early winter with wheat, barley, beans, onions, flax and clover. Summer crops could be grown only where water could be pumped up from the river by hand. The system was crude and although the ancient Egyptians developed one of the most advanced and enduring civilizations the world has ever seen, by the end of the eighteenth century AD, Egypt was in serious trouble. High floods left water stagnant in the basins where they provided breeding grounds for disease organisms. Low floods meant famine. By 1820 the population of Egypt had fallen to 2.5 million and the cultivated land area amounted to only 3 million acres. Then a series of low dams was constructed to hold back the river and channel it into canals. Water was available all the year round and farmers were no longer dependent on the annual flood. Summer crops, including cotton, maize, rice and sugar cane were introduced in addition to the winter crops. By 1955–6 the cultivated area had increased to 5.7 million acres and production had increased by 177 per cent.

The masters of irrigation, however, are the Indians and the Chinese. Writing of their achievements, Prof. Borgstrom says: *

'On a small as well as a big scale the ancient Chinese and Indian civilizations were pioneers in conquering the climate along this front. And throughout history China and India have maintained the lead in irrigation. They possess, even in this era of huge dams, more than half the irrigated acreage on the globe. During this century and even in the fifties they have carried out irrigation projects which in toto amount to some of the largest undertakings in history.'

Three-fifths of the irrigated land in Europe is in Spain, says Borgstrom, but the remainder of Europe has an irrigated acreage only one-sixtieth that of China's. He mentions the additional area irrigated in China during the fifties. This came to 100 million acres,

* *Too Many*, by Georg Borgstrom.

or the equivalent of the whole of the cultivated area of Canada. Two-fifths of all the irrigated land in the world is in China. In India there are three irrigation projects, each of them larger than the Panama Canal in terms of the volume of earth removed in their construction. Indeed, the name 'India' means the land of the river, the river being the Indus. The Aryans, who arrived in India some 3,300 years ago, discovered the remains of an ancient civilization on the banks of the largest river they had ever seen. They gave the river no name, just 'Indus' – 'the river.'

Nevertheless, it may be that the limits have been reached for irrigation by this method, for although it may appear to be remarkable in terms of increases in total cultivable areas, there are grim dangers.

Frequently, the area to be irrigated is flat and poorly drained. The water reaches the fields and the surplus, the water that does not evaporate immediately and that is not taken up by plants, sinks into the ground. There it goes to swell the underground water reserve, the ground water that moves slowly towards the sea, far more slowly than the rivers above it. As the volume of groundwater increases, so the level – the water table – rises. The water table may rise high enough eventually to waterlog the surface. The Indus flows through what is today West Pakistan and in parts of West Pakistan waterlogging is a serious problem. It can be solved, of course, by draining the land, but this is expensive. Today the fact is that although the Indus and its five tributaries, together with the flat plain across which they flow, are one of the major resources of the planet, the 30 million people who live there, 23 million of them making their living from agriculture, are poor and ill-fed. The plain is poorly drained and the irrigation canals are leaky. At one time the ground became waterlogged only close to the rivers. Now waterlogging affects wide areas.

Nor is waterlogging the only problem. As the ground water rises, it brings with it salts, dissolved from the subsoil. As the water reaches the surface, the pure water is distilled out and the salts remain in the top soil. Slowly the soil becomes more and more salty until it is rendered completely sterile. It is desert. Again, this problem can be solved. The water table must be lowered. This means efficient drains must be installed. Then, when the water itself is under control, irrigation must be resumed to re-dissolve the salts and flush them out of the soil. It is a long and expensive operation, but it can be done. To cure the problem in West Pakistan at the present time would mean treating 5 million acres, which is about 18 per cent of the total area under cultivation. It is ironic that agricultural development in West Pakistan is held back by waterlogging and salinity in some areas and by water shortage in others. Irrigation is concerned with no more than the redistribution of available water, but it can be more complex than it seems.

The Middle Eastern 'fertile crescent' between the Tigris and the Euphrates once fed the civilizations of Mesopotamia and Babylon. It may well have been waterlogging and salinity that caused the deterioration of what was, in its time, some of the best soil in the world. Great damage was done when waves of nomadic invaders destroyed the canal system, but they may have only accelerated a process that was already well advanced. Today, only about one-third of the total 7 million acres in the crescent can be cultivated each year. The remainder must be left fallow and un-irrigated in order to dry out the subsoil.

There are some areas in which canal irrigation might help. The harnessing of the flow of the Tigris-Euphrates system could perhaps increase the cultivable area from 7 to 10 or 12 million acres. The Soviet Union is a rich country and can afford to develop its water resources. In parts of the Niger basin, in West Africa, in the Rufiji basin in Tanganyika, near Lake Kyoga in Uganda, in the monsoon regions of eastern India, through Burma and south-east Asia, and in the lower Ganges and Brahmaputra basins in East Pakistan and India, irrigation might increase agricultural productivity. The capital is lacking. It can come only from the developed countries.

Although apparently dramatic increases in productivity seem within our grasp if only the distribution of water could be rationalized in certain key areas, we must remember that whenever water is diverted from a deep flow to a shallow flow, or, to put it another way, whenever the surface area of a given volume of water is increased, evaporation increases. In a hot, dry climate this can represent a significant loss, so that although we may calculate the flow of a deep river, we cannot use the whole of the water we divert into shallow canals. It may be that by using it in this way we are diminishing the resource too quickly.

Israel has discovered one solution to the problem of evaporation. Water for irrigation in some areas is conveyed through sealed pipes, rather than canals, is then stored in closed tanks, is carried to the fields in a further set of sealed pipes and then the pipes are run beneath the surface of the soil in the area under cultivation. The water reaches the soil from below, through small nozzles in the pipes, released in the root area of the crop plants. Of course, the system is very expensive, much more expensive than open canals, but plastic pipes are available in large quantities, they deteriorate slowly if at all, and the system has the added advantage that fertilizers can be added to the water in the tanks in measured amounts and so can be delivered to the crop plants in a way that is far more controlled than is possible even under the agricultural systems used in Europe and America. Israel is a rich developed country, in UN terms, and its solution is not relevant to the developing countries as a whole, but it does illustrate the fact that, like most of the problems facing the world today, evaporation and salinity from irrigation systems can be overcome.

It was believed some years ago that the problem was not being tackled on a large enough scale. The water was there, it was not being brought under control. We were not thinking big enough. In 1960, the Mexican Government announced its plans for a major irrigation scheme that would increase the cultivable area by more than 2.5 million acres. The project would require the building of a number of large dams. The Mexicans have been successful. Food production has increased by an average 5.4 per cent per year from 1952 to 1969, although the rate of increase has fallen off more recently. During the same period the population has increased by a steady 3.4 per cent per year. So, for the time being, food production outstrips population growth. The advantage must be temporary, of course. The gain in productivity made possible by improved irrigation has been realized. It has given all that it will give. Once the 2.5 million acres are under cultivation, they will remain 2.5 million acres. Large though that figure may be, it is static. Food production has increased arithmetically. Meanwhile, population continues to grow exponentially. The Malthusian situation reappears. It reappears in exactly the same way, to diminish into insignificance any increase in cultivable land we may envisage.

The world's largest and most publicized dam is the Egyptian Aswan High Dam. Work began on its construction in 1960 and it is to be fully operational in 1975. According to the original plan, it will increase Egypt's farmland by 1.3 million acres – an area about the size of New York State, or the whole of the county of Norfolk in England. This represents an increase of 15 per cent in Egypt's total cultivable land area. During the time it will have taken to build it, the population of Egypt will have increased by 35 per cent. The gain will have been overtaken before the project can be completed. In point of fact, soil studies carried out in 1965 showed that Egypt has barely half this area of potential agricultural land to reclaim. At present it is estimated that less than 300,000 acres have been brought into production.

The Aswan Dam was highly political in its day. The Dam was designed by German engineers and the design was approved by the World Bank. The United States agreed to finance the construction, but subsequently it withdrew for political reasons, leaving the field to the Soviet Union, which, in turn, handed it back to President Nasser, who spent 1,000 million dollars on it. In the course of the wrangling, Egypt nationalized the Suez Canal, which caused the Suez War and allowed the Soviet Union access to Egypt and the Mediterranean. About one-third of the total cost – some 360 million dollars – has been spent on reclaiming the 300,000 acres of land that have been added to Egypt's farm area. This comes to 1,200 dollars per acre!

Is the Dam worth the price? Perhaps it is, if it means more people can be fed. Unfortunately, the financial cost is only a part, and a small part at that, of the total bill Egypt may have to pay in the long run.

At one time the Nile used to flood, bringing not only water, but silt to the plains. The silt was rich topsoil, washed down from the upper reaches. Previous irrigation schemes for the river, which involved controlling and channelling the floodwaters, permitted the silt to reach the agricultural land. The Dam, however, holds back all the water and, with it, the silt, which settles to the bottom. The water that is released contains no silt. Thus, while the Egyptian farmers may have water, they are denied one of the richest fertilizers there is. Not surprisingly, since no substitute has been found, the soil is deteriorating and will continue to deteriorate unless and until plant nutrients can be restored. Soil scientists have calculated that fertilizers containing the main plant nutrients – nitrogen, phosphate and potash – as well as calcium and magnesium, will be needed in the Delta within four or five years. The soil of the Delta was once considered inexhaustible. Within fifteen to twenty years there will be serious deficiencies of copper, zinc, molybdenum, boron and manganese. Where are the nutrients? They are collecting in the lake at the rate of 134 million tons a year. The Dam is silting up, as all large dams silt up.

At the same time, the flow of water into the Delta and so into the sea, has been reduced. Salinity has been a problem in the Delta for a long time, but the annual flood used to wash out the salts. Now they are accumulating. The water that does reach the sea is saltier and contains fewer nutrients. This damages the soils of the Delta, but it also affects the sea itself. We have seen already how the concentration of pollutants in a river may be increased when water is abstracted upstream. This increased concentration of pollutants – for the Nile is as much of a sewer as any other major river – also reaches the sea. The water at the eastern end of the Mediterranean now contains between 38.7 and 39.8 parts per thousand of salt, which makes it saltier than the Atlantic, which has 35 parts per thousand of salt, and it is dirtier. The Egyptian sardine fisheries have been all but wiped out. In 1965 the catch was 18,000 tons. In 1968 it was 500 tons.

The silt-free water that flows from the Dam erodes the banks and undermines barriers and bridges down the whole of its remaining course. The cost of building control dams downstream of the main Dam is almost half that of building the High Dam itself.

Lake Nasser, the artificial lake behind the Dam, spreads Nile water over a larger surface area than it would have covered had it remained within its banks. This has increased evaporation. Of course, a proportion of the floodwater would also have evaporated, but this evaporation occurs upstream, before the water has reached agricultural land, so it has no opportunity of benefiting crops. Combined with seepage into the ground below and around the lake, the loss from Lake Nasser is now so great that Egypt actually has less water than it had before the Dam was built.

For centuries the Egyptians living close to the Nile have been exposed to a debilitating disease called bilharzia, or schistosomiasis. The disease is transmitted by an aquatic snail and during the dry season the population of snails was reduced as they dried out and died. Now there is no dry season and the only check to the spread of schistosomiasis has been removed. Eventually, it may affect the entire population.

All in all, the Aswan High Dam, planned to be one of the wonders of the world, may turn out to be one of the world's worst and most expensive ecological catastrophes. Conceived on a grand scale and built at vast cost to bring the standard of living of the Egyptians up to that of Europeans, within a single generation they are worse off than before the Dam was built, and entirely as a result of it.

It is interesting to note that at a meeting of the Institute of Water Engineers held in London in 1960, shortly before work began on the construction of the Dam, an Egyptian hydrologist, Dr. Abdel Aziz Ahmed, warned of many of the dangers that lay ahead. All of his prophecies have been born out, and other problems have occurred that he did not foresee. He returned to Cairo immediately after the meeting and he was fired. So much for Dr. Ahmed.

The story of the Aswan High Dam can be repeated, on a smaller scale, for many of the world's major dams, and there is, in addition, reason to suppose that in some areas large dams may be related causally to the incidence of earthquakes.

If we have gone almost as far as we can go by way of increasing the efficiency of irrigation, can we increase the overall amount of freshwater available? What about the icecaps? 97.74 per cent of the fresh water in the world is trapped in the polar icecaps. Hope springs eternal in the technologists' breasts and it has been suggested that we tow icebergs south (or north) and melt them where there are water shortages. This is pure science fiction, as is the proposal that we use atomic reactors to generate heat which we use to melt the icecaps. Should this happen, the fresh water would be released into the sea, so becoming salt and useless, while the change in salinity would be likely to alter the ecology of the oceans in unpredictable ways that would be likely to affect fisheries. The level of the seas would rise throughout the world by some 50m. A rise of 50m. in sea levels would flood New York and London as well as many of our other major cities. Holland would disappear completely! Included in the land that would be lost would be large areas of what is now highly fertile farm land, while the fraction of land that would be reclaimed would, generally, be much poorer. Moreover, the melting of the icecaps would produce major climatic changes. At present the large areas of white snow and ice at the poles reflect incoming solar heat and light back into space. If they were removed, the surface would be warmed by radiation and the air above it by convection. An increase in the temperature of the air would result in a general rise in world temperatures. The climate would grow warmer. Those regions that suffer from scorching heat would scorch even more and remain just as dry as they are now, while the total area of hot, dry desert would increase.

No, there is nothing we can do to add significantly to the world's stock of fresh water and there is little we can do even to distribute it more efficiently. While industry is the main consumer of water in the developed countries, throughout the world as a whole agriculture uses far more than any other human activity. It takes 120 to 150 gallons to produce one egg, 1,000 gallons to produce a quart of milk, 3,500 gallons to produce a pound of beef and 300 gallons to produce a pound of bread. If agricultural productivity is to increase, one of the first requirements is additional water, and there is little more water.

It would be unwise to belittle the seriousness of the water shortages the world may be facing. On 13 April 1971, *The Times* carried a report from Michael Knipe in New York describing a drought that threatens to stop agriculture in Texas altogether:

'Vast stretches of Texas and Florida are undergoing an extreme drought, the worst in 15 years. The survival of crops, cattle and wildlife is seriously threatened and with no end in sight the situation is being compared to the great drought of 1934, which caused the eroded areas of the Great Plains to be termed the "dust bowl" of America.

'In some parts of Texas there has been no appreciable rain in seven months and 60 counties have been proclaimed disaster areas by the Governor, Mr. Preston Smith. In the south of Florida some 6,000 brush fires have burnt over nearly 400,000 acres since the beginning of the year.'

He goes on to report that there were twenty-three dust storms in Texas since the beginning of the year, 'four of them violent whirlwinds covering hundreds of miles.' These dust storms may be the first of many, for there is a chance, based on past records, that Texas may be in for dry weather for the next five years. Most of the wheat crop was written off for lack of water to germinate the seeds. In Florida water was being pumped into the famous Everglades National Park to enable wildlife to survive.

In Britain, the Dee Weather Radar Project is developing a system of water conservation using radar and computers. A small radar unit scans clouds to determine the water content – this can vary considerably. The radar readings are fed into a computer together with information on the moisture content of the soil, the amount of vegetation on the surface and other data, all of which can be used to calculate the amount of water that will run off when it rains. This information in turn can be used to regulate the sluice gates within each catchment area which control the flow into reservoirs and rivers. The aim is to reduce the quantity of water that reaches the sea.

Britain is a wet country, yet even it has water problems. In the arid areas of the world the picture is grimmer.

At the 56th session of the FAO Council, held in Rome in June 1971, Mr. Brahim Aissa, a delegate from Algeria, appealed for help to stop the advance of the Sahara into Algeria and Upper Volta. He was supported by Mr. Léopold Wantissé Siry, of Upper Volta and Prince Mohammed Al Faisal, of Saudi Arabia. The encroachment of the desert into the cultivated areas of Algeria was, said Mr. Aissa, causing serious concern in his country. The Forestry Department of FAO was being asked to consider the possibility of re-forestation. Mr. Siry said that 'The slow but continuous advance of the desert in the areas bordering the Sahara is also a matter of great concern to our government. Many cultivated lands are being lost each year to the desert. As a result this may raise in the near future a problem of displaced population.' He warned that 'the problem may take on dramatic proportions if urgent measures are not taken.' Saudi Arabia, too, suffers from the problem of moving sand dunes that cause the disappearance of grazing areas.

It is in this world context that the agricultural productivity of the third world must increase. It is in this situation that attempts are being made to introduce the new cereal varieties, which require more water than the traditional varieties. The world shortage of water is immediate and, in the long run, it is insoluble. The best we can do is to buy time.

Time is valuable. Time can be used to bring other problems closer to solutions. But time costs money. There is a limit to the improvement that can be made in irrigation, but some improvement is possible. It requires capital. Capital can come only from the developed countries.

5

Fertilizers

After water, the next major requirement for the expansion of agricultural production through the use of the new hybrid cereals is fertilizer. Consideration of fertilizer use cannot be divorced from that of water, for the effectiveness of modern fertilizers is closely related to the availability of water in the soil. Shortage of water will inhibit the growth of plants, but before that point is reached it will reduce the efficiency of fertilizers and in some cases it will make their use positively dangerous both to the plant and to its consumer. We have considered already the problems associated with increasing the availability of water, but it is worth bearing in mind that limitations in water supplies will affect food production in these two ways.

The concept underlying the use of modern chemical fertilizers began more than a century ago. Almost since the beginning of agriculture itself farmers have known that growing plants removes substances from the soil that must be replaced. At one stage in the history of agriculture the method by which the lost nutrients were restored involved abandoning the cropped land as soon as yields began to fall and moving to fresh land. The land that was left was colonized by wild plants until it was green once more. After a lapse of several years it would be ploughed and planted again and yields would be as high as ever. Farmers knew that in some way the resting of the land restored its fertility. Then they noticed that plants grew larger, and more quickly, in the vicinity of animal droppings and near the carcases of dead animals. They began to collect dung and to apply it to the land from which they hoped to produce food. Blood was used, and bone. It is said that the early European settlers in North America found that the Indians increased the size of the maize plants that they grew by burying a fish with each seed. Other substances have been used: in Britanny farmers have used seaweed for centuries. By the nineteenth century, British farmers were using farmyard manure, pulverized bones and animal wastes, as well as guano – bird droppings imported from Peru.

Yet no one knew why manures and fertilizers had the effect that they did. Chemists became intrigued with the problem and they succeeded in establishing that nitrogen was a common constituent of many of the substances in common use. Nitrogen became recognized as a prime requirement for growing plants. The largest part of the air we breath consists of gaseous nitrogen, but it is unreactive. If it is a requirement of plants, and if their growth can be stimulated by it, then how does it arrive in the soil under natural conditions? It was a long time before the answer to this question was discovered and its discovery was related to another observed phenomenon. Certain farm crops seemed to restore the soil in a manner similar to manures. In time it was discovered that these plants – legumes – have tiny nodules attached to their roots that contain micro-organisms that are able to utilize atmospheric nitrogen and by forming compounds of nitrogen they make it available to their host plant. When the plant is harvested the roots remain in the soil, together with their nitrogen-fixing bacteria and the soil nitrogen content is increased. More recently it was found that other soil bacteria perform a similar function so that today we know that despite the relatively large quantities of nitrogen fertilizer applied by some farmers, by far the largest part of the nitrogen available to plants in the soil is obtained by the fixation of atmospheric nitrogen by the soil micro-population. The importance of soil micro-organisms in the nitrogen cycle extends even beyond this, for even the nitrogen fertilizer applied by the farmer cannot be taken up by plants until it has been 'processed' by certain of these organisms.

This is recent knowledge, however. By early in this century the first of the chemical fertilizers, by-products from industrial processes, were becoming available. Ammonium sulphate, a by-product of coal-gas works, could provide ammonia and Britain was importing sodium nitrate from Chile for use in explosives. The introduction of these chemical alternatives to traditional manures was rapid. In 1900, 91 per cent of the nitrogen applied to farm crops was in an organic form – a product of living processes. By 1913, this proportion had fallen to 40 per cent.

In 1910, a German chemist, Fritz Haber, designed a process for combining hydrogen with nitrogen to form ammonia. By 1913 the first fertilizer factory as such was producing 20 tons of ammonia a day.

Nitrogen was not the only plant nutrient whose essential nature was recognized. Not all the traditional fertilizers had nitrogen as their principal constituent. Bones contain large amounts of phosphorous. The connexion between bones and phosphorous was discovered in 1769 by a Swedish scientist, Johan Gottlieb Gahn, but it was not until 1840 that Justus von Liebig suggested that the phosphorous in bones could be made available to plants if sulphuric acid were added to them. Von Liebig is often regarded as the founder of the modern fertilizer industry and of the theories of plant nutrition upon which it is based. He believed that the nutrients essential to plants were three in number and that they could be supplied chemically, so removing totally and forever the need for any organically-based manure or fertilizer. There is a somewhat apocryphal story that on his death-bed he recanted and repented of his arrogance in assuming that he could isolate and identify the processes by which God made plants grow.

Others were not long in following up von Liebig's work with phosphates. John Bennet Lawes, a wealthy British farmer, patented a sulphuric acid-bone ash process in 1842 and so produced the first chemical phosphate fertilizer. His farm, at Rothamsted, is now one of Britain's leading government agricultural research stations. By the 1860s Britain was producing something like 200,000 tons of commercial phosphate fertilizer a year.

The third major plant nutrient to be identified was wood ash – called 'pot-ash.' The relevant constituent of potash is potassium and by early in the nineteenth century potassium chloride deposits were being mined in Germany and France to make fertilizers. The first commercial potash fertilizer factory was built in 1861.

The main advantage of chemical fertilizers is that they are less bulky than their organic counterparts and their nutrient content can be guaranteed during manufacture. Generally, they are applied in a granular form or sprayed in solution. In either case, because they are water-soluble their nutrient content is available to plants more rapidly than insoluble, or less soluble, organically based alternatives.

Until recently, it was believed that once dissolved these salts were taken up directly by plants. Since their nutritive value was known with great accuracy, the concentration of nutrients could be increased to far beyond what could be available in the soil naturally or in a less easily available organic form. So plants were stimulated and grew larger and more rapidly than would be possible under the most advantageous natural conditions.

There can be no doubt whatever that the introduction of chemical fertilizers increased food production, but it is impossible to know accurately by what amount. Many soils are naturally deficient in one or more nutrients and the application of fertilizers to such soils will restore them to a satisfactory level. Production will increase, but whether it has increased beyond the level it would reach had the soil not been deficient in the first place is difficult to tell. Crop yields have risen, nevertheless. In 1940 the British national average yield of wheat was 18.6cwt per acre. In 1968–9 it was 28.2cwt per acre. Barley yielded 17.6cwt per acre in 1940 and 27.4cwt per acre in 1968–9. Oats increased from 16.5cwt per acre in 1940 to 25.6cwt per acre in 1968–9. Yet during this same period which saw the most rapid increase in fertilizer consumption, there were also major developments in plant breeding. New crop varieties were introduced that produced heavier yields. No one can know precisely what proportion of the credit for the increase in agricultural production should be ascribed to increased fertilizer use and what to the improved varieties. There are farmers who do not use chemical fertilizers at all and the level of their yields, although it is often below that of the farmers who do use fertilizers, suggests that the major yield increases may be due more to plant breeding than to fertilizer use. If this is so, it would suggest that the fertilizers are replacing nutrients extracted from the soil by cropping but that they are adding little beyond this.

The three main plant nutrients are usually referred to simply as N, P and K. Consumption of NPK is increasing all the time. In 1968–9, the south-east region of England received an average total of 210.56lb of fertilizer per acre of arable land (99.68lb N, 54.88lb P and 56.00lb K). The eastern region, Britain's main cereal growing area, received even more – 228.48lb per acre of arable land (91.84lb N, 59.36lb P and 77.28lb K). Over the country as a whole, between

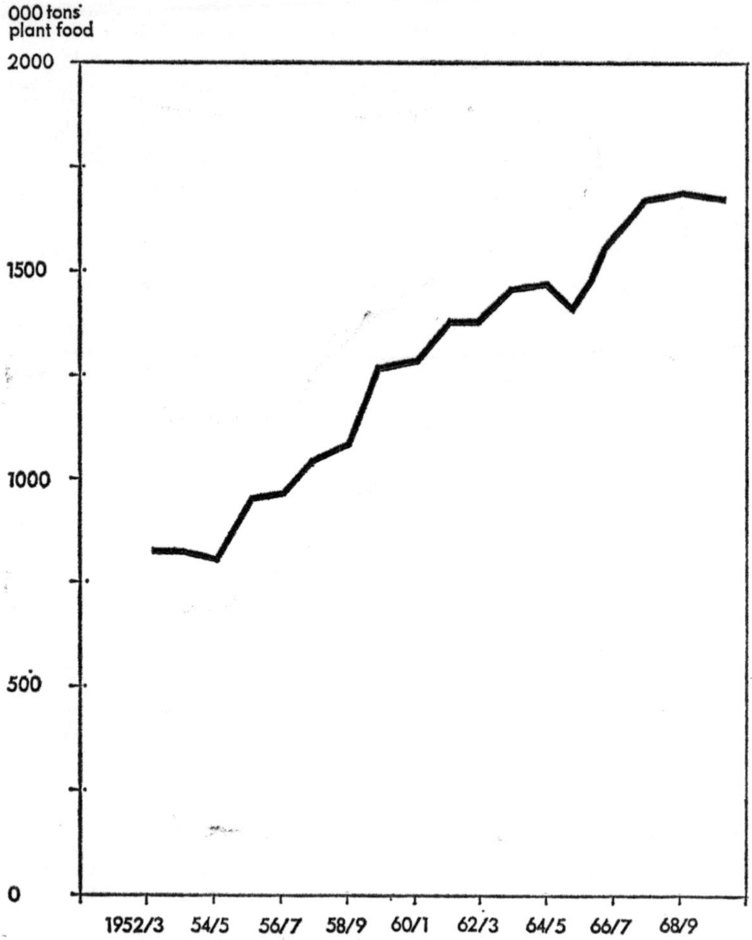

UK fertilizer consumption

1957–8 and 1968–9, fertilizer consumption has increased by 150 per cent in the case of N, 41 per cent in the case of P and 32 per cent in the case of K. Consumption continues to rise.

Obviously, there is a limit. A plant will grow only to a certain size no matter what inputs are applied to it. Once this maximum size is reached, no amount of fertilizer will make it grow larger. It is the aim of many farmers to achieve this ultimate in production. The nutrient most associated with growth is nitrogen and it has been found experimentally that crop yields will continue to increase with applications of N up to about 200lb per acre. Yields begin to level off after about 120lb per acre, but it should be possible, in theory at least, to continue up to 200lb before the diminished return from increases in input becomes serious enough to impose an economic limitation. In fact, in the state of Washington, yields have gone on increasing up to over 300lb per acre of N.

The theory seldom holds good in practice, because fertilizer is only one of the inputs required. Plants also need sunlight and water and no matter how much fertilizer he applies, a farmer will not grow a maximum yield if the weather is consistently poor. Agricultural scientists have had to admit that in the end farming depends to an unsatisfactorily large extent on luck!

There is an optimum level of fertilizer application and when this

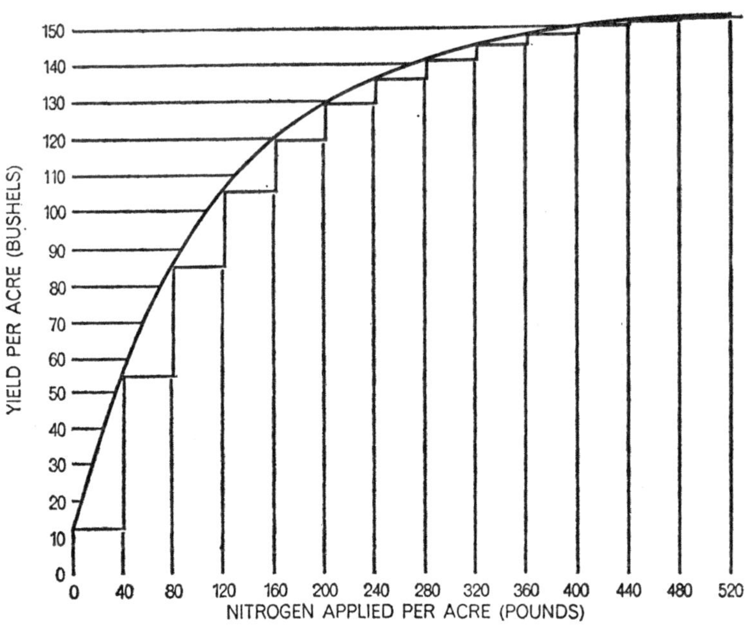

The economics of fertilizer use. As fertilizer application rates increase, resultant increases in yield decline. This graph is based on a crop of irrigated corn grown in the state of Washington, USA.

is exceeded things begin to go wrong. If the input of nitrogen to a cereal crop exceeds this optimum, the effect may be to so increase the size of the ear that the plant stem cannot support the weight and the crop 'lodges' or falls down. This makes harvesting difficult, or impossible. It is almost impossible to forecast what this optimum nitrogen application will be, because the fertilizer is applied at the start of the growing season and its effect is regulated by the water available after application. Since irrigation is uncommon for cereal crops in temperate climates, the water availability is related directly to rainfall, and this is not predictable very far in advance with any accuracy. So variations in rainfall will affect the crop. Too much nitrogen may also result in poorly filled ears and low yields. If there is heavy rain, the farmer may have applied too much fertilizer. If the weather is dry he may have applied too little. In the case of root crops, nitrogen appears to act in a similar way to irrigation. Leaf and haulm growth is stimulated and the formation of tubers is delayed, although in the end yields are usually higher. If still more nitrogen is applied, leaf and haulm growth is stimulated still further and yields are reduced. The additional plant tissue is generated in the leaves.

There are other dangers in the over-use of fertilizers. In sugar beet, for example, the whole pattern of amino acids may be altered, with a consequent change in the quality and balance of the protein content of the crop.

The addition of large, concentrated doses of particular substances to the soil may produce subtle chemical changes in the plant, by altering the balance between the quantities of the major nutrients compared with those of certain other mineral elements that are present in fertile soils and which plants require in very small amounts. Since the overall quantities of molecules that enter the roots of a plant is controlled, a relatively larger intake of some may be associated with a reduced intake of others. The plant might show signs of nutritional deficiency, but even if no visible sign were apparent, changes in its chemical structure might alter its nutritive value as a food.

The basic nitrogen fertilizer is ammonia. If ammonia is treated with sulphuric acid, ammonium nitrate is produced and this, too, is a useful fertilizer in common use. If too much nitrate fertilizer is applied to certain crops it may lodge, unmetabolized, in the plant tissue. This can have very grave consequences. Nitrate-nitrogen fertilizers have been used for many years. In addition to ammonium nitrate, there are others, such as potassium nitrate, ionic nitrate or, occasionally, sodium nitrate.

In the 1930s it was suspected that there might be a link between nitrate lodged in plant tissues and human and animal health. During the following decade the mechanism was identified by which animals were sometimes poisoned if they grazed pasture that had been over-fertilized with nitrates. By the 1960s similar mechanisms had been identified in humans and there were indications that at sub-lethal levels, nitrate could have an adverse effect on health.

The problem, it was discovered, was not the nitrate at all, but nitrite. It was known that nitrate in foodstuffs broke down into nitrite in the presence of water in the rumen of animals. Then it was discovered that this reduction was due to bacterial action. Nitrite is able to pass through the gut wall and is absorbed into the bloodstream. There it forms a stable compound with haemoglobin, forming methaemoglobin. This effectively locks up the haemoglobin so that it cannot combine with oxygen in the lungs to form the unstable compound oxyhaemoglobin. So the supply of oxygen to the body is reduced and in extreme cases it can lead to death by asphyxiation. The condition is known as methaemoglobinaemia. A further by-product of the microbial reduction of nitrate to nitrite is hydroxylamine, which is also able to enter the bloodstream directly through the gut wall and which plays a principal rôle in the destruction of vitamin A as well as having a haemolytic action which can give rise to anaemia.

In this way the poisoning of cattle and sheep was explained. Then it was found that human infants have in their gut the microbes that convert nitrate into nitrite. The deaths of babies from methaemoglobinaemia as a result of eating plant products which had received overdoses of nitrate fertilizer have been reported from Germany.

To some extent the reduction of nitrate to nitrite occurs at random in the environment. There is concern that a proportion of this nitrite may combine with amines to form nitrosamines. Since the amines are present naturally, and since the nitrate-nitrite is being released in uncontrolled amounts, the effect on the total environmental nitrosamine load is unknown. It is important. Nitrosamines are recognized as among the most potent carcinogens – cancer-causing substances – that are known.

Of course, we are talking here of applications of nitrate fertilizer – the criticism might not apply to other sources of nitrogen – that are too high. There is good reason to believe that nitrogen fertilizers are over-used by some farmers in Britain and in other developed countries, but this is hardly a risk in the developing countries. While in East Anglia the cereal farmers were applying an average of 230lb of fertilizer per acre, the arable farmers of the third world averaged 11½lb per acre, or 13kg per hectare.

The new varieties of cereals are defined as those which give a linear response to fertilizer inputs up to 70 to 90lb per acre. It is not likely that the arable farmer in the third world will encounter many of the problems of over-fertilization at these levels of use, but there is reason to suppose that if yields were to be maintained, these rates of application would have to be increased with time. We will discuss these reasons in more detail when we examine the effect on soils of intensification in chapter 7. Even then, however, it is unlikely that fertilizer consumption will reach European, American or, highest of all, Japanese levels. Not, that is, in terms of fertilizer per crop, but one of the advantages of the new varieties is their short growing cycle which makes it possible to grow two or even three crops within one season. Thus the fertilizer applied per acre per year – and these are the figures quoted for British farming – would be double or treble those per crop. 180 and 270lb per acre are levels comparable to those consumed in the developed countries and if applications were to increase in order to maintain yield levels they would exceed those of the developed countries. There is no doubt that the undesirable side-effects experienced in Europe and America would be experienced in the developing countries. Indeed, as we will see in chapter 7, they would be likely to be worse.

Will it happen? Obviously, the fertilizer industry hopes it will. In 1966–7, the developed countries produced 86 per cent and consumed 71 per cent of the world's total output of nitrogen fertilizer. During that year world production was 22,200 million metric tons. It is estimated that total world production in 1970 exceeded 30 million metric tons. By relating production capacity, population growth, food requirement and past trends in consumption, it has been estimated by the chemical industry that by 1975–6 the developed countries may have increased their consumption by 11.4 per cent (exponentially, of course) from the 1966–7 total of 15,810 thousand metric tons to 41,700, while the developing countries may have increased theirs by 10.6 per cent, from 6,390 thousand metric tons to 15,800. This gives a total world production in 1975–6 of 57,500 thousand metric tons.

Can the world produce this much fertilizer? The agrochemical industry is concerned at the present time, not with its ability to produce the fertilizers so much as with its ability to sell them. It is faced with gluts and over-production.

Pure nitrogen gas is the principal constituent of air. So long as the planet has air it will have nitrogen. The basic, and perfectly satis-

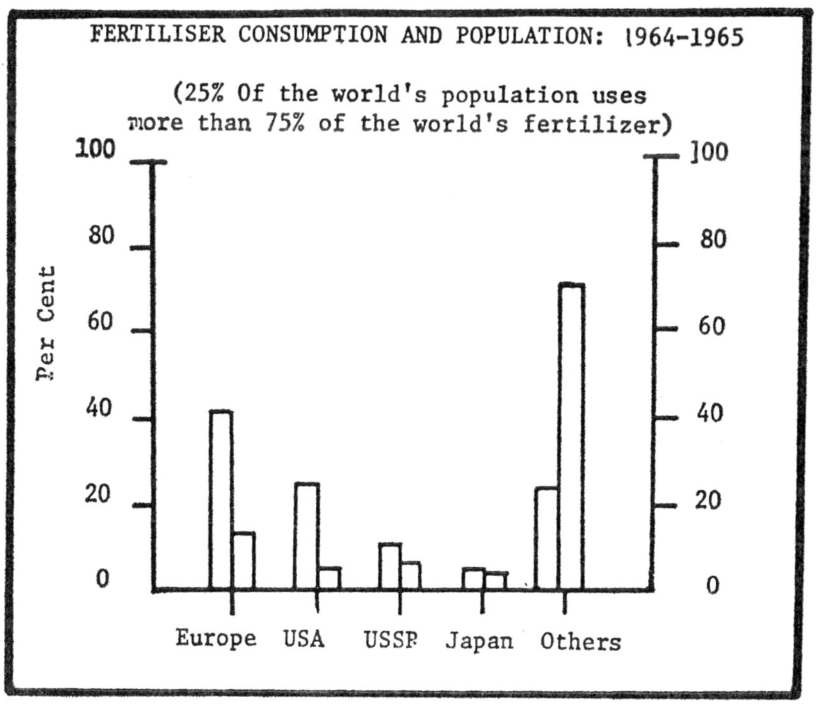

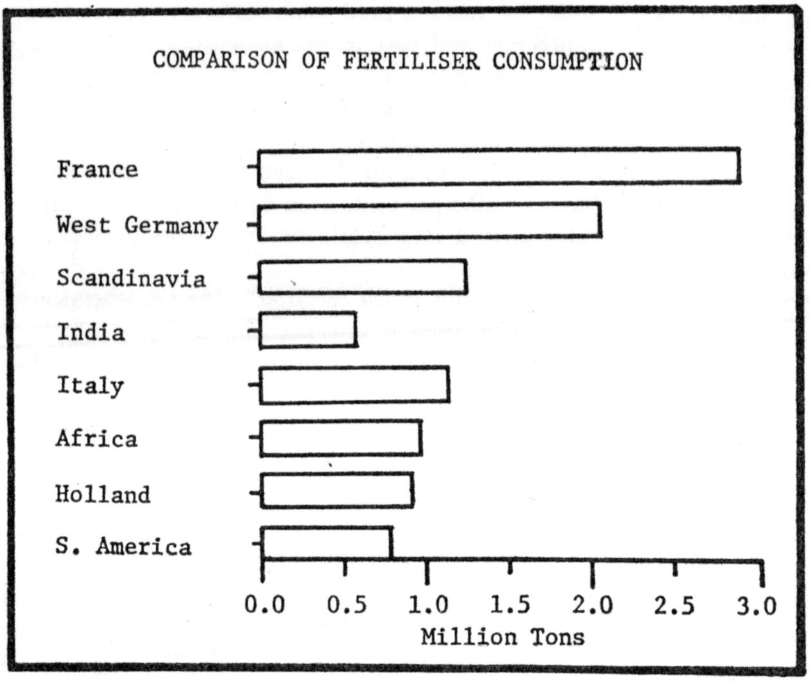

45

Fertilizer consumption and population, 1964–5. 75 per cent of the world's fertilizer production is used by 25 per cent of the world's population.

factory, nitrogen fertilizer is ammonia, NH_3, a compound of nitrogen and hydrogen. Hydrogen is a constituent of water. So for as long as the planet has water there will be hydrogen. The hydrogen can be extracted from the water by electrolysis and when the two gases are brought together under pressure (of at least 3,000lb per sq. in) in the presence of a catalyst (osmium) ammonia is produced. Sea water is as good as, and perhaps better than, fresh water as a source of hydrogen. So the basic requirements for nitrogen fertilizer are air, water and power.

On an industrial scale, it has been the provision of the gases in sufficiently large quantities at an economic price that has provided most of the serious development problems. Although it is possible to derive hydrogen from water, this process is little used. At the present time it is cheaper to obtain it from natural gas or from naphtha, which is a by-product of the refining of petroleum. In Britain, ICI's fertilizer production is being converted from a naphtha to a natural gas process. Global reserves of both natural gas and petroleum are limited, but provided the means exist to generate power, the production of ammonia will remain technically simple.

There is a further possibility that may affect nitrogen fertilizer production quite radically. At the University of Sussex, scientists are attempting to harness and cultivate the nitrogen-fixing bacteria that provide most of the nitrogen available naturally in soils. Should they succeed in isolating, cultivating and harvesting cultures of these organisms, it may be possible to manufacture nitrogen fertilizer biologically, in the soil itself, in the way nature does it. Whether this will prove practicable and economic remains to be seen, but at least it is a possibility.

Nevertheless, there are limitations to the production of nitrogen fertilizer and it would be naive to consider only the availability of the raw materials in global terms and leave it at that. As we have seen already, the world's stock of fresh water is limited and it would seem a self-defeating operation to deprive agriculture of water in order to manufacture fertilizer! In the long term, therefore, if water is to be used as a source of hydrogen, it is likely to be sea water. Immediately this restricts the location of fertilizer factories to coastal areas. The price of fertilizer will increase further inland and countries that are completely land-locked will be at a disadvantage.

Even without this restriction, and with other sources of hydrogen, the siting of fertilizer factories poses problems. In the developed countries, fertilizer factories tend to be very large and have extremely high rates of production. This is necessary if the price to the farmer of the finished commodity is to be kept acceptably low. In general, industrialists in the developed countries tend to 'think big.' As the technology is exported, the outlook is exported with it and technologists in developing countries tend also to 'think big.' Also, developing countries regard fertilizer factories as major prestige projects. They have acquired an almost mystical significance, related as they are to food production. So they are usually large, like their counterparts in the rich countries. Yet it is often the case that the poor country lacks the industrial and technological facilities to back up projects of this size and lacks the distributive mechanism to ensure that the product reaches the farmer, who may live thousands of miles away.

This tendency to 'think big' also affects the power input requirement. One large factory may consume less power per ton of product than ten small factories, but the provision of power in smaller amounts may be much simpler. In the developed countries power is generated relatively cheaply and plentifully at the present time from the burning of fossil fuels or from nuclear reactors, but a substantial reduction in the power requirement might make other sources adequate for particular purposes. You cannot generate very much electricity from a water wheel in a river, or from a windmill, but it might be possible to power a small factory from a number of such devices connected in series. It may be that the production of large amounts of nitrogen fertilizer will remain beyond the capability of many of the developing countries until they devise small, low-powered plants scattered, presumably, along the coast. Indeed, as the availability of power from fossil fuels and nuclear energy declines in the rich countries, they, too, may be forced to adopt similar processes and accept a reduced fertilizer output. When the power requirement is taken into account, the apparent ready availability of nitrogen and hydrogen appears in perspective and the possibility of producing more than 50 million tons of nitrogen fertilizer a year becomes debatable.

Phosphate fertilizers (P) begin as mineral calcium phosphate, which is mined. Opinions vary about how the deposits originated. Some geologists hold that they were precipitated from seawater after it had become saturated with phosphate and fluorine ions derived from the contact of the water with igneous rocks and gases. Others believe it may originate from the replacement of calcium carbonate by calcium phosphate in particles of a mineral called aragonite, which is found on ocean floors. If this is so, the process may be continuous and deposits may be being laid now that could provide future phosphate stocks.

Today phosphate deposits are found in areas vacated by the sea. As these areas have weathered, other substances which once cemented together particles of phosphate have been removed, leaving the phosphate in the form of small pellets. It is these pellets that are mined. They are insoluble in water and as a result they have been scattered widely during the course of time. It is a long time: phosphate deposits were being laid down 400 million years ago.

Once mined, the pellets are cleaned and pulverized to give rock phosphate. This is used as a fertilizer. Although it is insoluble, and so not available to plants directly, it reacts with acids to produce a phosphate that is soluble, while at the same time tending to neutralize the acidity of the soil.

The simplest chemical treatment of the evaporite involves mixing it with sulphuric acid. This forms a slurry which dries into a hard pile, which is ground again. It is called 'black phosphate' because of its colour or, more usually, 'superphosphate.' This can be, and is, marketed as a fertilizer. It contains about 18 per cent of water-soluble phosphorous pentoxide. It may be used separately or mixed with other fertilizers. Superphosphate may also be used as a raw material from which pure phosphorous can be obtained for making matches and for use in animal feeds.

The next stage involves treating the insoluble phosphate with a mixture of sulphuric acid and phosphoric acid. This produces gypsum and additional phosphoric acid, which can be separated by filtration. The gypsum is discarded and the phosphoric acid is mixed with more rock phosphate to form a slurry again. The slurry hardens and when it is ground it will contain about 48 per cent of soluble phosphorous pentoxide. This product is marketed as 'triple superphosphate.' If, instead of phosphoric acid, nitric acid is used to treat the insoluble rock phosphate and the insoluble phosphate content of superphosphate, the resultant fertilizer will contain both phosphate and nitrogen.

The material produced by acid treatment – the 'wet' process – is crude and contains many impurities. If a purer product is needed there is an alternative 'dry' process, in which the rock phosphate is burned in a furnace and the ash dissolved.

Both wet and dry processes produce satisfactory fertilizers and it is economics that dictates which process will be used. At present the dry process is not employed in Britain, simply because the wet process is cheaper. Where pure phosphorous is required in Britain it is imported from Newfoundland.

Rock phosphate, superphosphate and triple superphosphate can all be used as a fertilizer. The difference between them relates only to their water-solubility and, therefore, to the speed with which they can be made available to the growing plant.

There is no immediate shortage of rock phosphate, although we are mining it more rapidly than it is being redeposited. If power is available to provide heat, it can be converted into a fertilizer without the addition of any other raw material. Without power, sulphuric acid will be required. Does the world have sufficient sulphur?

In October 1968, the Society of Chemical Industry held a symposium in London to discuss the availability of plant nutrient reserves. S. L. Levitsky, of the Texas Gulf Sulphur Company, reported that the availability of sulphur had moved from a period of shortage in 1967 to one of abundance in 1968. It had been estimated, he said, that the world demand for sulphur would reach 32 million tons by 1970, which was 6½ million tons more than was produced in 1967. However, he went on, because sulphur enters into the production of so many products apart from fertilizers, demand can be linked to population growth.

'In the Free World as a whole, the average per capita consumption of sulphur was 23.5lb in 1965. Annual growth rate during the 1960s was 3.8 per cent annually, compared with an average 3.0 per cent over a period of 20 years. Even on the basis of this lower growth rate of 3.0 per cent a year, sulphur producers have to be prepared to produce 30 million tons of sulphur a year by 1970, and over 50 million tons by 1980, on the basis of the most conservative United Nations' population forecasts.'

Is there enough sulphur? At present the main reserves are located in France, Mexico and western Canada, with Canada being the world's leading producer. Reserves in Alberta are estimated at nearly 100 million tons and mining technology improves all the time, thus extending the mineable reserves. The search for sulphur deposits that can be mined economically has extended to all continents, but present production and the main part of the search, relates to elemental sulphur. If sulphur compounds are taken into account the reserves are much greater, but so will be the extraction cost. According to Levitsky:

'The size of these reserves makes these forms of sulphur eminently suited to become a permanent source of supply for decades to come. But the cost of extracting and transporting sulphur from such sources is high, requiring heavy investment on the part of producer and consumer alike.'

In the immediate future, therefore, it seems that the wet process may remain more economic for the manufacture of phosphate fertilizers, but that further ahead – Levitsky spoke of satisfactory, but expensive, sources of sulphur 'for decades' rather than centuries – the dry process may become the cheaper.

Neither will be cheap for the developing countries unless, as with the production of ammonia, they are able to design small, scattered factories powered by electricity generated locally at low cost.

Potash is also mined, and at present manufacturers are facing problems of over-production. World consumption in 1970–1 is estimated to be more than 18 million tons and by the 1980s this consumption is expected to increase to some 30 million tons. At the forecast demand of 22 million tons in 1974–5, present known reserves are sufficient for another 600 years. We must remember, however, that if consumption is to be related to increasing food production, related in turn to growing population, then the exponential reserve index will be considerably lower than 600 years.

There are various techniques for the processing of potash fertilizer, but the simplest involves pumping water into the potash beds, bringing the solution to the surface and drying out the solid potash. A number of mining companies have concessions to mine under the North Yorkshire Moors in England, where annual production could soon be equivalent to about twice Britain's annual consumption. The Yorkshire reserves are not being mined because at present there is no world demand for more potash.

The main resource requirement for potash fertilizer production, therefore, apart from the reserves themselves, is water, and as we have seen the availability of water may become critical in the years immediately ahead. Production is linked to the source of supply of the potash reserves. The main deposits are located in West Germany, East Germany, France, Spain, the USSR, Israel, USA, Chile, China, Italy, France and the UK, and there are further deposits, which are not being mined at present, in Denmark, the Netherlands, Poland, Rumania, India, Australia, Mexico, Brazil, Peru, Canada, Morocco, Libya, Tunisia, Congo, Ethiopia, Jordan and Pakistan. Altogether it is estimated that the world reserves at present amount to 123,864 million metric tons.

It is difficult to obtain estimates of potash demand very far ahead, although the FAO considers that in 1985 there may be a world demand for 30 million metric tons. In the near future it is assumed that the developing countries will not increase significantly their share in the world demand. Remembering that demand beyond about 1975 is not being considered, the OECD has concluded:

'The progressive exhaustion of certain deposits may in the long run create a shortage in some countries' home production, affecting the international import and export flow of potassic minerals. The substantial increase in output in other countries will also affect the international pattern of trade in K_2O. On the whole, however, the ratio between probable output and consumption does not seem to indicate any strain on the supply of K_2O.'

Not, at any rate, before 1975!

What is actually happening at the present time in the developing countries? Are they managing to obtain the fertilizers they require? In 1968–9, the total world production of fertilizers amounted to 61.3 million metric tons, of which 56.7 million metric tons was produced in the developed countries. In the same year, total world consumption amounted to 59.3 million metric tons, of which 48.8 million metric tons was consumed in the developed countries. These figures suggest a current surplus in fertilizer production in the developed countries that is not being taken up by imports into the developing countries.

Jordan's Crown Prince Hassan visited Europe in August 1971, seeking economic aid to develop his country's fertilizer industry, based on phosphates. The aim of the development would be to bring Jordan's phosphate production up to 3 million tons a year by 1975, much of which would be exported. The Jordanian reserves are large, but inaccessible without heavy foreign investment.

Kuwait, which bases its economy on petrochemicals – for the time being! – began work in 1966 on a complex of four units which would produce 400 tons a day of liquid ammonia, 550 tons of urea, 500 tons of ammonium sulphate and 400 tons of sulphuric acid. The scheme ran into a long series of difficulties and it was not in full production until 1971 nor, until that year, was the managing company able to pay interest and make a profit. So far no dividends have been paid. If Kuwait manages to solve its immediate problems, it is faced with competition from Iran, Iraq, Qatar and Saudi Arabia, all of which are entering nitrogen fertilizer production. The commercial future is not promising. Beyond the immediate future, the petroleum reserves will begin to exhaust.

Herculean efforts have been made to establish a viable fertilizer industry in India and in May 1971, the Indian government-owned Fertilizer Corporation announced proudly that it would double its production within two years and increase it by five times by the mid 1970s. By 1974 work should be completed on the construction of a urea manufacturing plant in Kalol, Gujerat, with a daily output of 1,200 metric tons. This, says the British company that has won the contract to build it, will be larger than any other single stream urea plant in India.

In India, however, the fertilizer industry is not healthy. S. C. Anantharaman reported in the *Financial Times* on 25 January 1971 that:

'The fertilizer industry in India is in serious trouble. New capacity is not coming up according to target; production is still around 50 per cent of capacity already created; and consumption is not picking up as expected in spite of the increase in overall food production.'

The FAO has a programme which aims to increase production and availability of fertilizers in developing countries. On 2 February 1971, FAO reported that between 1961 and 1970, recipient countries have increased from six to twenty-two, while the number of field experts has increased from nine to fifty-five. A total of thirty-four countries have benefited from the programme so far.

The Fertilizer Programme is financed jointly by the fertilizer industry and by donor and recipient governments, but it operates on a very low budget. By February 1971, 1.5 million US dollars, in cash and fertilizers, had been pledged for the coming year. Of the donor governments, West Germany had guaranteed 575,000 dollars, the UK 300,000 dollars, Belgium 60,000 dollars, Finland 48,000 dollars, Algeria 50,000 dollars, Brazil 40,000 dollars and Indonesia 17,500 dollars.

In 1969–70, Europe produced 25,630,000 metric tons of fertilizer and consumed 23,390,000 tons, the USSR produced 9,820,000 tons and consumed 8,040,000 tons, North and Central America produced 20,010,000 tons and consumed 16,580,000 tons. In the same year, South America produced 420,000 tons and consumed 1,230,000 tons, Asia produced 5,790,000 tons and consumed 7,320,000 tons, mainland China produced 1,690,000 tons and consumed 3,250,000 tons, Africa produced 1,160,000 tons and consumed 1,470, 000 tons and Oceania produced 1,300,000 tons and consumed 190,000 tons.

It seems, therefore, that for the next few years at least the IWP fertilizer requirement for developing countries can be met only if domestic production is supplemented by large scale imports from the developed countries, paid for with aid channelled through the Fertilizer Programme. At present the demand is not being met, the Fertilizer Programme is short of funds and the blame must lie with the developed countries.

In the middle term the aim will be to increase fertilizer production in the developing countries themselves. In the case of potash there will have to be international trade since potash can be obtained only from the countries which possess the reserves. In the case of nitrogen and phosphates the raw materials are adequate but it is questionable whether developing countries possess, or will possess, the capital or the technological and industrial capability to establish successful fertilizer industries on the scale that will be required. We should remember that the aim is to bring fertilizer consumption in the developing countries up to levels comparable with those in the developed countries.

In the long term the availability of certain of the mineral resources is questionable and the availability of water and cheap power may prove limiting factors.

It is the aim of the IWP to increase the fertilizer consumption of the average arable farmer in the developing countries from 11½lb per acre to between 90 and 270lb per acre by 1985. This would appear to be hopelessly ambitious.

6

Pesticides

Left to itself, an area of natural vegetation will contain as many different species of plant and animal as it will support. It is a basic law of biology that if there is an ecological niche it will be filled. Should any one species increase in numbers so as to unbalance the system, there is a series of checks by which its numbers will be reduced and the balance restored. A population increase of a particular species will be followed by population increases of those species which eat or parasitize it. Once it reaches its climax, an ecosystem will remain stable until some outside factor intervenes to disturb it.

The aim of agriculture is to disturb the natural system, to adapt it to man's advantage. The farmer does this by clearing away the natural vegetation so as to expose land which he sows to a limited number of species. This represents a simplification of the system. In terms of balance, and the diversity of species that existed before he interfered, the farmer has created a situation in which one species – his crop – preponderates. The inevitable reaction must be for the species of animal that graze the crop plant, the micro-organisms that parasitize it and the plants that can compete with it for sunlight, water and nutrient to multiply in accordance with the natural laws of balance that operate within the system. As the other species increase in number, so they in turn will attract 'pests,' 'weeds' and 'parasites.' A complex network of relationships will spring up, all tending to displace the crop plant and, left to itself, the natural system will recover. This is what is meant by 'the balance of nature.'

For his part the farmer must outwit the system, but he has to reckon with weeds, pests and diseases. They are inevitable. The story of man's battle with nature is, to a large extent, the story of agriculture.

Anything, any technique or substance that assists the farmer in this battle is to be welcomed. Yet there is a danger that in his attempts to overthrow the system he may so unbalance it that he can continue no longer in the face of its attempts to recover, attempts that will appear to him like massed onslaughts.

There are a number of ways in which the farmer may cheat the system, as it were. They form the basis of most good farming. But perhaps this is not enough. Perhaps more is needed. In Britain, not so many years ago, farmers believed that they should allow a certain proportion of their seed, or their crop, to be taken by pests, and that a certain loss from disease was inevitable. They allowed for it and their farming was successful, since they looked forward realistically to yields that took account of these losses. As the pressure on the farmers increased, as the need to grow more and more food was reflected in the agricultural support system of successive governments, they found it necessary to harvest what had been lost. The weeds, pests and diseases had to be eliminated so far as might be possible.

It is difficult to calculate precisely how much food was lost to pests and diseases. It is not easy to calculate such losses even today. According to the manufacturers of agricultural chemicals, some 60,000 acres of wheat in Britain are liable to attack each year from wheat bulb fly. Without seed dressings to prevent such attacks, they say, this pest would account for the equivalent of the crop from 30,000 acres. The British Ministry of Agriculture's National Agricultural Advisory Service (since disbanded and replaced by the Agricultural Development and Advisory Service) has calculated that over a twelve-year period frit fly, which attacks cereals, has caused a 16 per cent crop loss over the country as a whole. Throughout the whole range of plant crops produced in Britain it is estimated that the loss due to pests and diseases may amount to 10 per cent each year.

Britain, remember, is a country in which farmers have ready access to the most sophisticated materials and techniques for pest, weed and disease control.

As soon as man became aware of the fact that he could use poisonous substances to reduce the number of his competitors for food, he has used those substances. In the early days some of the simple inorganic compounds were very poisonous indeed. Many herbicides were based on arsenic and cyanide. The use of weedkiller as a murder weapon was a stock ingredient in the plot of countless detective stories written before and even during the Second World War. But big changes were coming. During the war, chemists had been preparing a whole new range of substances more deadly than anything the Borgias could have imagined.

One of the more sensational ones is called fluoroacetamide. It was sold widely as a rat poison. It achieved notoriety as a result of what came to be known as the Smarden Incident. Early in the 1960s a veterinary surgeon, Douglas Good, was asked to investigate the deaths of farm animals around Smarden, in Kent. Clearly, they had been poisoned, but his autopsies revealed no poison in their bodies. The deaths were a mystery. Then, walking one day along the tow-path of a canal, Mr. Good noticed that the plants were dead and dis-coloured. There was a factory nearby that he knew was engaged in the manufacture of fluoroacetamide, but the company assured him that a seepage from the works was impossible. Certainly no fluoro-acetamide had been detected in the carcases of the dead animals. It was not until biological tests were undertaken that the poison was identified. Meat from the poisoned animals was fed to dogs, which died and were examined very soon after death. Fluoroacetamide was identified as the agent, and it was discovered eventually that, through no fault of the manufacturing company, there had indeed been a seepage into the soil in which had grown the grass on which the farm animals had grazed. The carcases were burned and the contaminated topsoil was removed, sealed in canisters and dumped in the sea. It was the murderer's ideal weapon. It was colourless, odourless and tasteless. A minute dose resulted in swift and certain death and the poison then broke down in the body, leaving no trace. Fluoroacetamide was withdrawn from sale, although it is still used in some cities by local authorities and under very strict control.

This was a dramatic instance of an accident involving a modern pesticide. Not all are as dangerous as fluoroacetamide, but many are highly potent and must be handled with skill and care.

The first main group to be introduced after the war were the organochlorines, the most common example of which is DDT. All the organochlorines – or chlorinated hydrocarbons as they are sometimes known – are based on a complex molecule composed of carbon linked with hydrogen – a hydrocarbon. The simplest hydrocarbon is methane – marsh gas – which has one carbon atom and four atoms of hydrogen. If one of the hydrogen atoms is replaced by an atom of chlorine, we have the simplest of the chlorinated hydrocarbons, methyl chloride. Substitute three chlorine atoms for three of the hydrogen atoms and we have chloroform, and replace all four hydrogen atoms with chlorine and we have the common – and toxic – cleaning fluid, carbon tetrachloride. From this base it is possible to construct elaborate arrangements of atoms to form complex molecules that resemble closely many of those found in nature. Because of this similarity, the organochlorine compounds may enter the tissue of animals, but once there they cannot be metabolized in the usual ways and they interfere with a number of bodily processes, particularly with the central nervous system. As poisons they are extremely effective. DDT has been used for so long and on such a wide scale that we think of it as harmless. It is not, for, like other related compounds, it possesses a further quality: it is chemically stable and deteriorates only very slowly. The organochlorines are insoluble in water, but they are highly soluble in fats, so they tend to accumulate in fatty tissue. According to Rachel Carson, whose book *Silent Spring* brought the problems of modern pesticide use to the fore, 'Once it has entered the body it is stored largely in organs rich in fatty substances . . . such as the adrenals, testes or thyroid. Relatively large amounts are deposited in the liver, kidneys, and the fat of the large, protective mesenteries that enfold the intestines.

'The storage of DDT begins with the smallest conceivable intake of the chemical (which is present as residues on most foodstuffs) and continues until quite high levels are reached.'

We will consider pesticides as pollutants later, but at present it is worth noting that they are all potent and should be handled with great care.

The second main group are the organo-phosphorous compounds. These are not persistent; they break down quickly and this restricts the environmental damage they can do. However, some of them are, if anything, even more toxic than the organochlorines. They are cholinisterase inhibitors. Messages from motor nerves to muscles are conveyed chemically, by a substance called acetylcholine. This is produced by the body in minute quantities and it causes muscles to contract. As soon as the message has been conveyed, and the muscle has obeyed, acetylcholine is neutralized by an enzyme, cholinesterase, and it disappears. Indeed, it is a very transitory substance, as it needs to be. If a further substance can be introduced that will inhibit the action of the cholinesterase, then the acetylcholine will not be neutralized and the message to the muscle cannot be cancelled. The muscle will contract, but as soon as it begins to relax it will receive a second impulse from the acetylcholine and so it will contract again. The result will be tremors, leading to convulsions, leading to muscular spasms, followed by complete relaxation – paralysis. In a mammal, when paralysis reaches the diaphragm, death ensues. The whole process is quick and unpleasant. The organophosphorous compounds inhibit cholinesterase. Again, they are effective against pests, but they are dangerous and should be handled with extreme care.

The third main group are the carbamates. Less persistent than the organochlorines, they, too, are cholinesterase inhibitors.

It is reasonable that substances designed to kill animals should be of some danger to other animals. Herbicides – weedkillers – are less dangerous, but even with them accidents occur from time to time. Fungicides, too, are less toxic to animals than insecticides, but even with them there are some risks and a number of fungicides are based on highly toxic heavy metals, such as mercury.

The new cereal varieties, on which so many of the FAO's hopes are pinned, are particularly susceptible to attacks by pests and disease and it is considered necessary to increase the use of pesticides in developing countries. Leaving aside for the moment the risks from accident or from pollution, will they work and is it likely that they can be made available in the quantities required?

Modern pesticides were developed in the rich countries to deal with problems in those countries and for use by farmers in those countries. Their success is difficult to estimate, because it is all but impossible to assess with any accuracy the losses due to pests, weeds and disease. These losses were only part of a wider problem, however.

In the industrial countries it has been economically advantageous to divert labour from agriculture into urban-based factory industry.

This has been an historical process and it continues today. It has meant that farmers in the industrial countries have had to make do with a progressively smaller labour force. Therefore, they have had to find techniques that are economical of labour. The extent of the decline in the agricultural labour force is very great and it should not be underestimated. In the ten years from 1959 to 1969, for example, the full time labour force in the eastern region of England fell from 99,864 to 58,136; in the East Midlands it fell from 43,847 to 26,598; in the North from 30,923 to 19,083; in the South East from 73,082 to 42,843; in the South West from 62,019 to 36,778; in the West Midlands from 49,151 to 28,437; in Yorkshire and Lancashire from 38,794 to 22,956; and in Wales from 27,182 to 14,981. Throughout the whole of England and Wales, labour is leaving the land each year and with less than 2 per cent of its total working population employed in agriculture, forestry and fisheries taken together, Britain has one of the world's most efficient farming systems – in terms of output per man. A farm that employed twenty-seven men within living memory may now employ only three or four.

It is in this context that many of the changes in farming must be understood. The 'agricultural revolution' has been aimed only partly at increasing food production. It has also saved labour and this has been the underlying problem; not how to get rid of pests and weeds, but how to do so without employing men to do the job. When we attempt to translate agricultural techniques employed in Europe and the US into the systems of husbandry that are practised in developing countries we must beware of assuming too readily that the needs of one society are identical with those of another, or that what we consider to be improvements are necessarily related to total food production. Herbicides, for example, are labour-saving and nothing else. Weeds used to be removed from the fields by hand. When the hands are no longer available alternatives must be sought, but when they are available it makes no kind of sense not to use them.

The introduction of pesticides produced a further problem. For generations, farmers had had to rely on their own skill to defeat pests, weeds and diseases. They had to accept certain losses. Sometimes they might lose an entire crop. If they were to survive commercially, they had to farm in such a way that the disturbance they caused in the natural ecosystem was kept to a minimum. They had to offer the smallest target to invaders, by restricting the areas devoted to one species. They had small fields, with different crops in each. They did not dare to grow the same crop on the same land year after year, because this would have permitted an accumulation of competitors specific to that crop. So they devised crop rotations. The pest that preferred wheat might attack the wheat, but after the harvest it would die away. It might lie quiescent in the soil for a matter of months and if wheat were grown there again it would attack again. Against a totally different crop it would be harmless and helpless and it would die. The farmers ensured that predators – the beneficial insects, birds and animals – survived. They cared for their hedges, ditches and stands of trees, because it was there that the predators lived.

The introduction of chemical methods of pest control removed the need for a whole system of farming that had been geared to preserving some kind of ecological balance. Suddenly it was possible to grow the same crop in the same field year after year. Hedges and trees were no longer necessary, except as shelter for livestock. Then the livestock was removed into indoor units and the last use for hedges vanished. The Ministry of Agriculture pays a grant to farmers to help towards the cost of hedge removal. Farming was able to become more intensive and it did. This development – and it has taken place over a number of years – was encouraged actively by successive governments and, naturally, by the agrochemical industry. Yet it is based on a fundamental misunderstanding of the purpose for which pesticides had been developed in the first place. They were meant as aids to the existing cultural system. They offered the farmer a tool with which he could control situations that otherwise would get out of hand and that might cost him a severe loss. The original intent was not that the tool should take over completely from the entire system and become the main prop of a new system. This is what has happened wherever pesticides have been introduced on a large scale. It is no exaggeration to say that the tool has become a crutch. Of course, if the new system is superior to the old, everyone will benefit, but is it? The Ministry of Agriculture publishes an annual list of approved pesticides for use by British farmers and growers. The 1971 list contains the names of nearly 800 commercial products. Why so many?

The reasons are complex, and although in principle they apply to all the products involved, the situation is demonstrated most clearly in the case of insecticides.

A farmer suffers from an outbreak of insect pests. However, the pest insects are not the only ones in his fields. There are others, many of them innocuous, some which would assume pest proportions were it not for the fact that a rival species has flourished and is monopolizing the food supply, others that predate on the pest and potential pest species. This is the situation into which the farmer introduces an insecticide which is capable of killing any insect, of any species. Because they are most numerous, he will kill more of the pests than of any of the others and because it was this species that was most noticeable the reduction in its population will seem very satisfactory. The spray did a good job. Proportionately, however, he may kill more of the predators, because they are fewer in number. If he sprays too efficiently he may kill so many predators that their population is reduced to so low a level that it is no longer biologically viable and it may die out temporarily. He will not have killed all of the pests. Those that remain now benefit from an improvement in their environment. Their food is still abundantly available and there are fewer predators than there were. They thrive and very soon the situation is worse than it was before. On the other hand, the farmer may indeed so reduce the pest population that another species is able to take over the habitat and become dominant. So for one pest the farmer has succeeded in substituting another. He may have so timed his spraying that he killed off a large proportion of the pest population just as the predator population was expanding to deal with it. With its food supply suddenly reduced, the predator may become a pest itself, or it may turn its attention to another, beneficial species, and so create a new problem. Insecticides are, at best, very crude weapons to use in a delicate and highly complex situation and it is for this reason that biologists talk of upsetting ecological balances – disturbing the 'balance of nature,' if you like – with such concern.

In all probability the farmer will not spray once, but a number of times. The first time he sprayed a proportion of the pests survived. Some were on the underside of leaves, or sheltered in some other way, so that they did not come into contact with the insecticide.

Perhaps they will be less fortunate at the second spraying, or the third. There were others, however, that were sprayed and survived. Among these there will be some whose chitinous outer skin was too thick for the spray to penetrate. There will have been others that possessed the ability to de-toxify the chemical within their bodies. In either case they were naturally immune and their immunity was based on a physical characteristic that they had inherited. All at once this characteristic acquires a survival value. They will breed. Since the desired characteristic is transmitted genetically, it will be inherited by their offspring. Each time their habitat is sprayed, more of their susceptible rivals will die, while they will prosper, until a significant proportion of the total population acquires the new characteristic. This is evolution by selection of those strains of a species with the highest capacity to survive. It is an inevitable biological response of a numerically large population to the introduction of a hostile factor into its environment, and the rate at which it occurs is related to the concentration of the insecticide and the frequency with which it is used.

What should the farmer do? His best course is to abandon the insecticide altogether and allow the pest-predator relationship to stabilize. More usually he will increase the concentration and/or spray again. This will win him a temporary advantage but it will encourage the evolution of still more highly resistant pests and so the situation will be worse than ever. The next step is to change the formula and to use a different substance altogether. Again, this will achieve its immediate aim, but the gain will be temporary, because the insect population will acquire a resistance to any substance if it is exposed to enough of it for a long enough time. So the apparent technological solution is no solution at all in the long run, because it is based on an incorrect assessment of the problem. The pests are no more than a symptom. The problem is the farming system that encourages pests, and that goes on encouraging them until it is changed.

It has been estimated that there are now some 250 species of insect pests that are immune to most insecticides in common use. Some tropical disease control programmes, which are based on eliminating the insect vectors with insecticides, are seriously hampered by the 100 or more disease-carrying insects that are now pesticide resistant.

The failure of insecticides is illustrated by the attempts that have been made in Queensland, Australia, to control the cattle tick, *Boophilus annulatus microplus*. *Boophilus* was introduced into Australia in the late nineteenth century and, in common with many introduced pests, it has no natural control. By 1900 it had spread widely and was causing serious losses. Around 1920 the practice was to dip cattle in a solution of arsenite of soda. This was partly effective, but by 1940 the insects were showing signs of resistance. Then DDT was introduced and it remained in use until the residues of DDT surviving in produce necessitated its withdrawal. It was replaced by non-persistent organophosphorous compounds. Then, in the mid 1960s, the problem became much worse. Farmers found they were having to dip their cattle more frequently and for the first time ticks were encountered during the winter, when previously the cattle had been free. More definitely resistant strains were appearing and the Queensland government declared all-out war on the ticks. Dipping was intensified, the insecticides being paid for by the government, and tick-free quarantine areas were established. All seemed to be going well when the campaign stopped – probably because of the cost of the insecticides. At that time farmers were dipping at double strength every seventeen days. The frequency of dipping was reduced and then the ticks increased again. Soon farmers were having to dip at double strength every ten days and the kill rate was no more than 40 per cent. The chemicals used were being changed regularly, but resistance was appearing almost as soon as the new product came into use. The whole situation was out of control. One despairing farmer complained that doses of insecticide that were capable of killing the cattle did not harm the ticks.

It may be that the solution should have been to exchange the cattle for other breeds that are tick-resistant. There are such breeds in countries where the tick is indigenous because, over many generations, mammals may acquire resistance just as insects do. It might have been possible to introduce predators. Again, there are predators in the ticks' country of origin. Mynah birds and cattle egrets both eat ticks. Be that as it may, clearly the problem cannot be solved by the use of insecticides.

Similar resistance to herbicides is appearing among weeds, and since the target here is more closely related to the crop than an insect is, there must be some hazard to the crop itself. It has been known for some time that herbicides reduce crop yields and on 20 April 1971, *British Farmer and Stockbreeder* reported that the use of herbicides might be rendering crop plants more susceptible to disease. According to the report, all herbicides applied to cereals were implicated, and potatoes were also affected.

These limitations to the use and effectiveness of pesticides are not the fault of the products themselves, nor of the scientists who developed them. They cannot be remedied by developing more powerful, or even more selective, materials. There is a contradiction inherent in the concepts and theories that underlie their use. This is not to say they have no part to play in food production, only that to a large extent this role has been misunderstood. In a paper presented to a symposium on potential crop production in Britain, held in Aberystwyth in July 1969,* Professor S. H. Crowdy, of the Department of Botany, University of Southampton, summarized the situation:

'The effectiveness of current insecticides has led to an almost complete dependence on chemicals for pest control, and the advent of more efficient and specific fungicides may encourage the same tendency in the control of diseases. This dependence is dangerous, since any breakdown in the chemical treatment renders the crops vulnerable to attack; the risk of a total breakdown in protection is best avoided by developing systems of integrated control, reducing the use of chemicals to a minimum, and taking full advantage of biological factors to reduce pests and diseases. These systems of integrated control will require an intensive study of the biology of crops and their diseases and pests, and of the effect of chemicals on the whole biological system.'

It is the aim of the FAO to accelerate the introduction of chemical pesticides into the developing countries, and the aim is endorsed with enthusiasm by the industry that manufactures them. A paper issued by the manufacturers quotes the IWP as saying:

* *Potential Crop Production,* edited by P. F. Wareing and J. P. Cooper, Heinemann Educational Books Ltd., 1971.

'The spread of high-yielding varieties, the increased use of fertilizer on all crops, irrigation development and a high rate of multiple cropping will all tend to generate increased pest and disease problems and thus increase demand for crop protection chemicals.'

The paper goes on to eulogize on the benefits to be derived from the use of two Shell products, aldrin and dieldrin. Both these chemicals have been banned or severely restricted in most of the developed countries, including Britain, because of their extreme persistence and their danger to wildlife, to man and to biological systems generally. The paper says:

'The agricultural contributions of two of the Shell organochlorine products, aldrin and dieldrin, is mainly through their pre-eminent position as remarkably effective soil insecticides and seed dressings; dieldrin is also used as a foliage insecticide.

'The advantages they have as soil insecticides are:

1 A wide spectrum of control (which is needed because of the wide variety of soil pests.)
2 Sustained activity (which is required because the life of most soil pests is measured in months or years rather than in days or weeks, while the only opportunity of applying the product is often before or at sowing time.)
3 Low cost.
4 Ease of application.
5 No taint in crops.

'Aldrin and dieldrin protect well over 25 million acres of crops from soil pests every year outside the USA. The type of land typically treated might support four people per acre so that the equivalent of food for 100 million people is being protected.

'Dieldrin also offers a cheap and effective method of controlling locusts; it cost 13p to kill a million locusts (1 sq. mile of an average swarm contains 10 million locusts) which consume in one day crops sufficient to feed 500 million people.'

We will have more to say later on the effects of persistent pesticides on the health of those who consume products that have been treated with them.

We have arrived at a situation where the developed countries are seeking for ways to reduce their dependence on pesticides because of their limited efficiency and their undesirable side-effects, while at the same time urging the developing countries to adopt them on a large scale because they are so effective and safe!

In point of fact, where are the pesticides? According to Shell Chemicals, the total world retail value of pesticides amounts to around 3,000 million dollars a year. Of the total produced, 50 per cent is sold and used in North America, 20 per cent in Europe, 15 per cent in the Far East, 10 per cent in Latin America and 5 per cent in Africa. In other words, 70 per cent of all the pesticides in the world are used in the developed countries; and the remaining 30 per cent used in the developing countries includes the consumption of one developed country, Japan. Almost all pesticides are manufactured in the developed countries.

There is a reason for this. Most pesticides are based on organic molecules and the organic chemical industry is closely linked with the petrochemical industry. Pesticides, or most of them, are petroleum by-products. They can be produced only as part of a larger petrochemical complex and this requires heavy capital investment and a high degree of technological and industrial sophistication and an adequate industrial environment within which plant can be maintained and developed and skilled operatives trained. Unlike fertilizers, which conceivably could be produced in small quantities wherever the power is available, pesticides are products of big industry and it is unlikely that the present range could be produced outside this context.

They are expensive. Again according to Shell, insecticides can cost 5,000 dollars a ton as they leave the manufacturing plant. Freightage must be added to this price before they reach their destinations and the more remote the destinations are, the higher the retail price will be. Against this must be set the fact that they are not bulky. Shell estimates that the entire world output of insecticides, in their most highly concentrated form, probably amounts to no more than about 400,000 tons a year. All the same, and in spite of the claims for aldrin and dieldrin, the price is high.

All modern pesticides are the result of long and elaborate development programmes. The problem of resistance among pests has underlined the need for these programmes, for the useful life of many products is limited. Also, competition among manufacturers is keen. A new product will not have the market to itself for long before a rival appears to outshine it. The cost of the laboratory and field work involved in the development of a new pesticide may amount to a million pounds or more. This cost must be included in the sale price and it is the main reason for the expensiveness of the products. As it is, a miscalculation of the commercial life of a product, which is taken into account in spreading the development costs, can leave the manufacturer with a reduced profit. From his point of view it is highly desirable that the market should be increased.

At the same time, it is possible to conduct research and development programmes only where there is a pool of scientists and technicians, together with the ancillary support they need to function. There must be universities, government laboratories and experimental stations, laboratory equipment manufacturers and suppliers, all linked into and paid for by the industrial-commercial complex.

It is clear, therefore, that for the foreseeable future pesticides will continue to originate in and be manufactured in the developed countries, and that the industrial countries will sell them to the poor ones. At the present time the principal Shell products are manufactured in Holland, France, Colorado, New Jersey, Alabama, Texas, Louisiana, Australia and Mexico.

This being so, the developing countries must be able to afford to buy them. This means that aid must be provided by the developed countries.

If the developing countries must learn to live without pesticides, what of the rich countries? Can they continue to produce and consume most of 400,000 tons a year, a rate of production and consumption that is increasing exponentially at 10 per cent per year?

What are the requirements for pesticide production? First there is petroleum, as a source of most of our organic chemicals. As we saw earlier, there is likely to be no petroleum left by the end of the century, so all petroleum-based products will have to be phased out before about 1990. Before that happens, their price will rise steeply as competition for the remaining petroleum reserves stiffens. It is likely that petrochemicals will price themselves out of business before 1990. There are alternative sources of organic chemicals, particularly coal, but the power and water required are likely to prove excessively expensive.

Even the petroleum-based products require vast amounts of power and water. Shell's plant at Carrington, near Manchester, consumes some 50MW of electricity per day. Will cheap power continue to be available on this scale? Britain plans to substitute nuclear reactors for its present sources of electricity supply, but at its present rate of consumption, uranium is in short supply. The nuclear power programme is behind its schedule, it is far more expensive than was anticipated and it causes grave pollution problems. The fast breeder reactors which should take over from the present generation of reactors are even further behind. The breeder reactor manufactures its own fuel and so is extremely economical of non-renewable resources. However, it is very far from safe and of the seven in operation in the world at the present time (one of which, at Dounreay in Caithness, has been operational for only two months at the time of writing) three have had major accidents. At present, breeder reactors do not work and there is doubt about whether they will be ready in time to take over from the present reactors. There is talk, too, of thermonuclear fusion reactors. They would consume sea water as a fuel. So far none have been built and opinions differ as to whether they are feasible.

Should the power be available, however, what of the water? The Carrington plant utilizes 150 million gallons of water a day. It is true that much of this can be recycled, but whenever water is used there is some loss by evaporation. Water lost by evaporation is lost altogether from the fresh water stage of the hydrological cycle. True, it will fall again as precipitation, but except under very unusual circumstances it does not fall in the area from which it was abstracted. At Carrington the water and power are used to produce steam at more than a million pounds an hour.

At Carrington they do not even manufacture a finished product. The plant produces cyclopentadiene, a raw material used in the manufacture, in Holland, of dieldrin and endrin. Before the pesticide is ready for the market there are still more resources to be consumed. If a more thrifty use of resources is forced on the industrial nations then pesticides may be found to be expendable luxuries.

The FAO requires the countries of the third world to increase their consumption of pesticides if its 1985 targets are to be reached. Such an increase is inhibited at the present time by the lack of capital and foreign exchange with which to buy them. Although the manufacturers would welcome a development of new markets in the developing countries, their governments are reluctant to provide large sums of money which would, in effect, subsidize their own private industries.

The developing countries cannot manufacture pesticides in large quantities themselves, because they lack the industrial and economic structures necessary for the operation of one highly specialized and sophisticated part of a very large industry.

The development of such industries in developing countries is inhibited further by the imminent shortage of all the raw materials and resources they require. As we have seen, these shortages will certainly prevent further industrial growth in the developed countries and will necessitate large-scale reduction in the present levels of industrialization. It is unlikely that pesticides will continue to be produced at all for more than another fifteen to twenty years.

Meanwhile, although the manufacturers and the FAO are urging the developing countries to increase their consumption, the developed countries, which are experiencing the side-effects and the beginning of what promises to be the end of their effectiveness, are seeking to reduce their dependence on pesticides.

For these reasons, we may discard the use of pesticides as a significant factor in future food production.

7

The Grass roots

There is an English parish, near Huntingdon, that was once primeval, mixed, deciduous forest, as were most of the lowland areas of Britain.

The Romans built a road through it to the north. They called it Ermine Street. There were one or two small settlements beside the road.

In Saxon times the number of the settlements had increased and in addition to the road itself, there were tracks leading into the forest. The Saxons had begun to farm small areas.

Some time between then and the fourteenth century, 1 sq. mile in the south eastern corner of the parish became a royal hunting forest: managed woodland. It remained so until the seventeenth century. This was Cromwell's country and during the Commonwealth period that followed England's revolution, the hunting forest was 'nationalized.' By the latter part of the century much of it had been cleared of woodland and it was divided up into small fields, separated by hedgerows.

A hundred years later it was much the same. More of the woodland had been cleared, the original settlements had become villages, the tracks roads. Almost the whole parish was being farmed. At the time of Constable, it must have looked much like the landscapes he saw and painted, some 70 miles away. So it remained until after the Second World War.

The story of this parish is the story of England. Gradually, over several centuries, the forest was cleared and farmers moved in to manage the land. Britain has been farmed for many generations.

It has been farmed well. Each time trouble appeared a remedy was found that saved the land. It did not always save the people who lived on the land. The eighteenth and nineteenth century Enclosures solved the problems of over-grazing on the common lands, but they did so by excluding most of the rural population from their source of food and fuel. They became large estates owned by the rich. It was good for the land, but it was disastrously bad for most of the people.

Farmers learned that what they took from the soil they had to return. It was a difficult lesson, slowly learned. Not all farmers at all times have learned it. Much of what is now part of the North African Sahara was once the granary of Roman Europe. Its crops were harvested and sent to the cities; nothing was returned. Sicily fed Rome itself and Victor Hugo commented that 'the fertility of Sicily went down the sewers of Rome.' The farmers of the American Middle West farmed in this way, extractively. They grew crops until the land would grow crops no longer, then they moved on. There have been times and places at which this kind of shifting cultivation has worked, but it is a wasteful way to farm. Land that has been exhausted must be left for many years before it can be farmed again and so either the population is constantly on the move or it must occupy a large area at a low density of population. The Medieval fallow, where land was left to rest, was a relic of shifting cultivation and even then the land was saved from complete failure only by the breaking up of the manorial farming system. History records many instances of cities, and civilizations, that have failed because they failed their land, their land failed them, and they could no longer feed themselves. The farmers of the Middle West were unlucky. They produced local climatic changes, and a desert.

It is easy to forget that soils must be tamed. Ours were tamed so many centuries ago that although reason tells us that our domestic livestock have distant wild ancestors, reason tells us no such thing about the soil. Yet it is true. A virgin soil, cleared of forest for the first time, may be very rich. It may grow excellent crops. But after a few years they will falter and yields will decline. The plants have been tamed – they have been selected and bred to grow larger and faster than their wild ancestors – and their needs are greater. They take from the soil more nutrients than do wild plants. They will exhaust the soil that is not accustomed to them. The soil must be fed and trained. It must be assisted to provide itself with alternatives to the trees that bound it together and drained it, or the weather will erode it. It must be allowed to accumulate a balance of nutrients and of stability until it becomes a bank of fertility. This is what has happened to many of the best of British soils. They are banks of fertility. Or they were.

There is nothing intrinsically 'special' about our soils. There is no reason to suppose that originally they were any better than those of any other region. They have been farmed well, that is all. The farmers have been helped by the British climate. Complaining about the weather is a world-renowned British pastime, and with justification. Farmers complain more than anyone. It is true. The weather is often awful. But the climate is excellent. It is its very unpredictability, the fact that its sunshine and rainfall are spread evenly, that makes it so good. At the same time the summer days are longer than they are further south and although it is cooler, there is some reason to suppose that the extension of the hours of daylight is of greater benefit to growing plants than additional warmth – provided that it is warm enough for them to grow at all, which for most of the year it is. The winters are comparatively short and much warmer than they would be in our latitudes were it not for the insulation provided by the oceans and by the Gulf Stream. Yet temperatures in winter do fall below freezing point, but not too far below and not for too long. The water in the topsoil freezes and as it does so it expands and bursts open large lumps of earth. This reduces the work involved in reducing them to a fine tilth for sowing.

Farming is never easy, but temperate farming is the easiest there is. With a long tradition behind us, and a healthy credit in the fertility bank, it is possible to think of taking a few liberties, of spending some of our capital. Provided we have a good reason for doing so, and are not simply wasting it, this will do little or no harm. We can always pay it back if we wish. We must pay it back, whether we wish to or not, if we plan to continue to grow food.

It is farming in this kind of climate and under these sorts of conditions that has been studied most by agricultural scientists. It is productivity on these lands that has been increased. If increased production through intensification can work anywhere, it can work here. What has happened?

The widespread use of fertilizers and lime dates from the 1930s. Fertilizers were available long before that, of course, but their introduction was slow, largely because farmers were too poor to buy them. Then the government stepped in and awarded subsidies to farmers to help pay for fertilizers. During the war years fertilizer use was actively encouraged by the government. Then, after the war, the fertilizer industry, relieved of the need to provide nitrates for explosives, but ready to operate at maximum capacity because of the explosives they had been making, turned their attention fully on to the farmers. Fertilizer usage has increased every year since then. As we have seen already, between the years from 1957–8 and 1968–9, nitrogen usage has increased by 150 per cent, phosphate by 41 per cent and potash by 32 per cent.

Production has increased by an overall 35 per cent in twenty years. This is a large increase on the original output in terms of actual food produced, but it averages a linear growth of less than 2 per cent per year, so that today Britain still produces only about half of the temperate foods it consumes. If the population crisis in developing countries is to be overcome by increasing food production, then they will have to do a good deal better than that. Of course, British yields were comparatively high to start with. In 1946, British farmers grew an average 19.1cwt of wheat per acre; by 1969 this had increased to 28.2cwt. Barley yields increased over the same period from 17.8cwt per acre to 27.4, oats from 16.3 to 25.4, potatoes from 7.1 to 9.8 and sugar beet from 10.5 to 15.1. In 1946 there were in Britain 9,629,000 cattle and calves; in 1969 there were 12,374,000. In 1946 there were 1,955,000 pigs; in 1969 there were 7,078,000. There were 20,358,000 sheep and lambs in 1946; in 1969 there were 26,604,000. Poultry numbers increased from 67,117,000 in 1946 to 127,220,000 in 1969. The average milk yield from dairy herds rose from 544 to 815 gallons per cow per year and the average number of eggs per hen per year rose from 108 to 211.

During the whole of this period, the agricultural labour force has dwindled, from 976,000 in 1946 to 413,000 – less than half – in 1969, and it continues to fall.

The size of the agricultural labour force is important, for in an industrial country it is advantageous to divert labour from agriculture into industry. As we have seen already, many of the innovations that have been made in British farming have been designed rather to save labour than to increase production. There are two kinds of productivity in agriculture, productivity per man and productivity

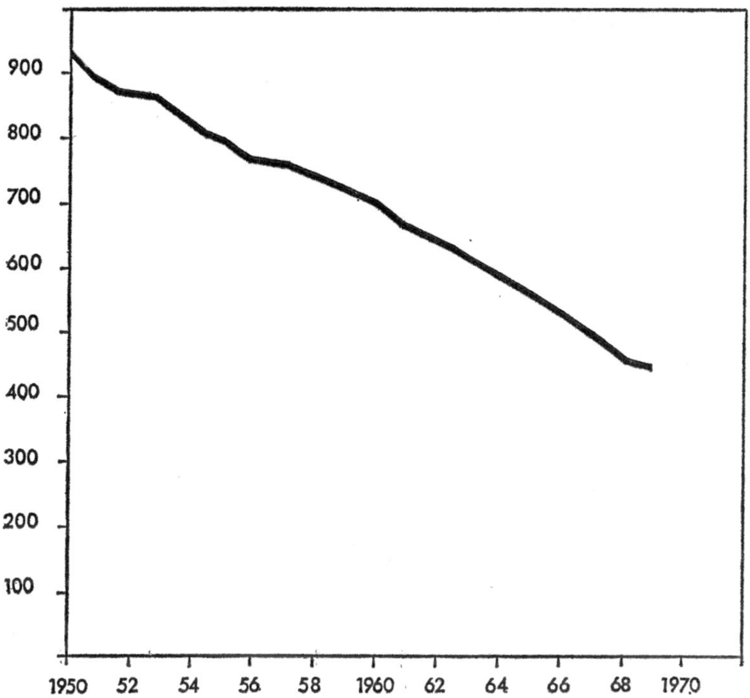

Number of agricultural workers in the UK. The figures are compiled in June each year.

per acre. The two are not necessarily related and it is quite possible to increase one without affecting the other, or to increase one and reduce the other. If a machine does the work of three and a half men, shall we say, productivity per man will increase by a large amount. We cannot have half men and so we will now employ one man where before we employed four. But because we have lost half the labour of one of the men, productivity per acre will decrease. The relevance of this to developing countries is obvious. Where there is rural and urban unemployment, tractors and combine harvesters may not help and total food output may be reduced.

So, with the aid of heavy fertilizer use, increasing mechanization and constant guidance from an elaborate network of government advisers, backed up by research stations, agricultural colleges and an industry, also with its own research facilities, providing the inputs, a comprehensive support system to subsidize the farmer and to guarantee him against losses, Britain has managed an increase of 35 per cent in two decades.

Is this the end? Will yields continue to increase indefinitely? The present evidence suggests that the yields of the major farm products in Britain has reached a maximum. Cereal crop yields have not risen significantly – in some cases they have not risen at all – since the early 1960s.

There is one obvious reason for this. A plant requires a number of inputs if it is to grow. It needs light, it needs an ambient temperature between certain limits, it needs water and it needs nutrients. Man has introduced drainage and irrigation to control the water economy of the soil, but he has not added to total water availability, nor has he gone very far, even in Britain, towards bringing water from the wetter areas into the drier ones. This has not been necessary, for nowhere is it so dry that crops cannot be grown at all. All that is necessary is to adapt the farming system so that each region grows the crops most suited to its climate. So the relatively dry east of England tends to grow arable crops, particularly cereals, while the relatively wet west country tends to grow grass and rear livestock. This is common sense, but it is also as far as we have been able to go in controlling the availability of water inputs. About sunlight we have done nothing at all. We cannot make the sun shine. It is possible to seed clouds and make them release their precipitation, but no practical way has been devised for moving the clouds along, out of the way. Cloud affects the intensity of sunlight and it also affects temperature, with which, again, we have done nothing. All that we have been able to do is to add to the nutrients that are available naturally in the soil and then to breed plants that are highly respsonsive to that one input. Probably this represents only a minor improvement over the natural situation. Although we have altered the status and, to some extent, the significance of one input, that is all that we have done, and it is all that we are likely to do.

In any case, no matter what it may be, there is a maximum size to which any plant will grow. Once it reaches that size it will grow no larger. Thus, the plant itself offers a limit to productivity. This is something we can do something about. The varieties of plants that are bred for their response to nutrient inputs may concentrate their additional tissue in those parts that are harvested. This may or may not be the whole plant. There is no way of telling how much of our increased productivity is due to fertilizers and how much to plant breeding, but the contribution of the plant breeders is certainly large. Even then, however, there is a limit beyond which further increases in fertilizer input produce adverse effects. With cereal crops the additional growth may be concentrated in the ear – which is desirable – but the ear may become too heavy for the stem. The plant may fall over – 'lodge' – and become difficult or impossible to harvest. At the same time, the enlarged ear may be poorly filled with grain. The yield falls. Judging the correct amount of nitrogen to apply calls for a considerable degree of skill and, like so many farming operations when it comes to it, a good deal of luck, for its effectiveness will depend on water availability and this, in turn, depends on rainfall. The fertilizer is applied at the beginning of the growing season. If the summer is too wet, the fertilizer applied in the spring may have been too much. If the summer is dry it may have been too little. Too much nitrogen applied to a root crop will increase the haulm growth. There will be more leaves above the surface, but not necessarily larger tubers beneath.

The breeding programmes that led to the new cereal varieties on which the 'Green Revolution' are based was searching for plants that would respond to applications of nitrogen. Plant breeding programmes in Britain are engaged in the same search. What is needed is a variety that will have a short stem and that will concentrate its growth in its seeds – the grain. The nutrients taken by plants from the soil are used to increase their overall size. Photosynthesis, which takes place in the leaves, produces carbohydrate and a proportion of this carbohydrate is stored in what will become the seed. The need is to alter the behaviour of the plant so that more carbohydrate is stored and so that green leaf will appear earlier in its growth, so extending the period during which photosynthesis can take place.

Will it happen? No one can say, but it has not happened yet. Unless and until it does, there seems little likelihood of further increases in yield from existing varieties. Fertilizer usage in Britain, certainly on cereal crops, is at its maximum and in places and at times it exceeds this maximum and the farmer suffers.

For the present, then, British food production appears to have gone as far as it can go. What is true of developing countries is true of the developed countries as well, and it is arable productivity that determines total food output. Livestock may be grazed, but the amount of grazing land is fixed. Animal productivity may be increased by breeding or by more efficient feeding. This may mean improved stock management, or improved grasses – where the criteria are different from those applied to cereals, but the problems no less intransigent – or the substitution of a more nutritious plant product. The more nutritious plant product is a grain and so we come back to cereal production. As we will see in a later chapter, the increases in livestock production in the developed countries is based, not wholly on an overall increase in their cereal output, but on the import of feedingstuffs, which is cheating!

Has British farming gone too far? There are some who say it has and there is reason to fear not only that yields have reached their peak, but that they may have passed it, and that the efforts of scientists and farmers in the immediate future will be devoted to sustaining the present levels.

Let us go back to our parish in Huntingdonshire. When we left it, in 1945, all the woodland had been cleared and the land was being farmed in small fields, surrounded by hedges. By 1970 a major change had taken place. All the hedges had been removed and the cropping area had become prairie. At the same time, the total land area had been reduced. Ermine Street had become the A1 and then a loop of the A1 was by-passed and the stretch of road through the parish became the A1(M). The Saxon settlements that became small villages now include an industrial estate and an overspill development. And there is a large US Air Force base. So the pressure on the land that remains has increased. This is one way in which population pressure is felt; more people need more food, but they also need land on which to live, so that the pressure on the farmer is felt from two directions.

During the winter the prairie is left bare and brown. There is nothing, no trees, no hedges, no plants of any kind, not even the stubble from last year's harvest, because that was burned. There is nothing to protect it from drying out as the wind whistles across the surface. There is nothing to shield it from the impact of the rain, which hammers it. The cattle that once grazed it have been removed long since. They are away on the other side of the country in an intensive indoor unit. So there is no dung, no residue from the food they ate to return to the soil to be ploughed back. The cycle is broken in the way that the Romans broke it in Sicily. Then it was the depletion of plant nutrients that caused the trouble. Today the problem is solved with fertilizers. Or is it? The structure of the soil is deteriorating. Will the fertilizers work when the soil is gone?

This is to state the situation in its most dramatic terms. The soil has not gone yet and it is far from inevitable that it will. Yet soil is a substance whose complexity is only partly understood and we know less than we thought we did about plant nutrition.

Soil is produced by the combination and interaction of two simultaneous processes. Bare rock weathers. The heat of the summer expands it, the cold of the winter contracts it and it cracks. The rain beats against it, dissolving out some of its water-soluble components to leave holes and gaps. The rain and the wind roll rocks back and forth, rock against rock, grinding them into small fragments. Each time a rock is broken its surface area is increased and the surface area is vulnerable to attack by certain primitive living organisms that can extract mineral nutrients from them. Colonies of micro-organisms establish themselves and, as they die, they provide nutrients for other organisms. Slowly the accumulation of finely ground mineral detritus and organic matter grows until it offers sustenance to plants. The blade of grass growing from a crack in a stone wall may seem to be rooted in the rock itself. In fact, it is anchored in holes in the rock but it is extracting nutrients from between five and seven generations of life forms that have gone before. Dying plants add to the organic matter, and all the time nutrients are being extracted from the mineral substrate, converted into the tissue of living organisms, and then broken down again into simpler forms to nourish a new generation.

So soil is alive. Something like 60 or 70 per cent of a good, fertile, agricultural soil is purely mineral, but the remainder includes about 8 per cent of living organisms. The topsoil of an acre of good farmland may contain 2.5 billion (million million) live creatures, including 26,000 earthworms. The weight of the living organisms in the top few inches of a field may exceed that of the cattle grazing the field.

There is a natural progression from the most primitive unicellular organisms to the mighty oak forest and all the insects, birds, reptiles and mammals that it sustains. In nature the succession leads to a stable ecosystem and it takes many thousands of years. When it is complete, when the system has developed to its climax, it will contain as many different species as can be supported. If there is an ecological niche, it will be filled. It is this diversity of species that gives the system its stability. Once established, there is no reason why such a system should not last forever, provided it is not disturbed.

Farming disturbs ecosystems. Indeed, it is one of its objectives. The farmer removes the natural vegetation in favour of particular plants that he has selected for their value to man. This simplification can be enough to make the system break down. Man did not always farm. As he evolved into the pre-existing system he filled a niche. For the greater part of his history he had little effect on the environment of the areas in which he lived. He ate what he could find – fruit, berries, roots, insects. He hunted and was hunted. Then he discovered fire and used it to clear the land he needed to grow the crops he was beginning to learn to sow.

Not all British soils have had such a fortunate history as those of Huntingdonshire. Professor Geoffrey Dimbleby, of London University, has studied pollen grains found on archaeological sites. For some reason pollen does not break down in the soil by bacterial action, and so he has been able to reconstruct the vegetation patterns of particular areas. He has made a special study of the heaths and moors of Yorkshire from the beginning of the Bronze Age. It was then that man began to burn the primeval forest.

When a forest is burned, the first plants to recover are grasses and shorter stemmed varieties. These provide good grazing for animals, but nature is resilient and left to itself the vegetation would recover. The trees would return and in the end the pattern would be much as before, although there would be rather fewer species. Since not all plants grow at the same rate, the more frequently a forest is burned the simpler the vegetation it will sustain. If, at the same time, the regenerating plant cover is over-grazed, the system will be simplified still further. When simplification reaches a particular point the sward will break up. As the countryside becomes more open the fire risk will increase. Winds will fan the flames further and more swiftly than they would in the shelter of the forest.

This is what happened in Yorkshire, until in the end the dominant species became heather, which is a perpetual fire hazard. The entire area is now semi-desert, like much of Highland Scotland.

It seems unlikely that we could repeat errors made in the Bronze Age, and yet modern intensive monocultural farming does represent a drastic simplification of the ecosystem. The small fields, surrounded by hedges, ditches and trees, included a large number of different plant and animal species and the fields themselves grew each a different crop and each a different crop every year. Once the livestock is removed, the hedges and trees gone and the crop rotation abolished in favour of the repeated growing of the same crop, the change in the ecology of the area is fundamental. There are obvious dangers. Such large areas devoted to one species are an open invitation to pests, diseases and weeds specific to that species, so that pest and disease problems are likely to be more serious than they were.

The removal of animals from the land has meant that their wastes – their dung – is no longer returned to the soil. At the same time, there is no longer any need for pasture, and so the grass has been ploughed up. Between 1964–5 and 1968–9, about a million acres of permanent or temporary grassland was ploughed up, while the acreage sown to barley increased by about 900,000 acres. It is an interesting fact that never in the history of British farming has so large an area been devoted exclusively to one crop as is devoted to barley today. Grass is unique among plants in its ability to stabilize soil structure, by breaking up large lumps and by binding the soil together within the intricate network of its root systems. It is well known in farming that continuous arable cropping will bring about a deterioration in structure, while stock farming or mixed arable and stock farming will maintain or improve it. At one time crop residues were ploughed back into the land. This continues, but on a reduced scale and straw stubble, the residue from cereal crops, is often burned. Agricultural chemists maintain that because there is a natural balance between nitrogen and carbon in soils, as in plants, the decomposition of the straw, which is mainly carbon, by bacterial action will consume nitrogen that otherwise would be available to the succeeding crop. At the same time the burning has little effect on soil micropopulations – the temperature of the top few inches is not increased very much and the fires are of short duration – while the ash provides potash and a number of trace minerals. Therefore, it can be argued that straw burning is good for the land. Nevertheless, it deprives the soil of a little more of the organic matter that would have contributed to its structure.

Farmers became concerned about the deterioration of soil structure several years ago. Mr. H. R. Fell, in a paper delivered to the Farmers' Club in 1967, drew attention to a deterioration in standards of husbandry and the effect it was having on soils. In 1969, the National Farmers' Union initiated an enquiry among its members in England and Wales to determine the extent of the trouble. In December 1969, the then Minister of Agriculture, Mr. Cledwyn Hughes, asked the Agricultural Advisory Council to undertake a full enquiry into modern farming and its effects on soils.

The report of the Council, entitled *Modern Farming and the Soil*, was published 12 January 1971. It found that soil structure was deteriorating over large areas of the country and that the organic matter content of arable soils was declining. The situation is being exacerbated by the use of heavy machinery, which can have a very deleterious effect on soil by compacting it, particularly if it is used when the soil is too wet. In some mixed farming areas there is overstocking of livestock, and on many farms the drainage systems are poor or inadequate. There are soils, such as the calcareous soils and rendzinas that cover the chalk and limestone of, for example, the Downs, the Cotswolds, the Chilterns, the Wolds and the magnesian limestone belt that runs from North East England south to Nottingham, which are highly stable. The mineral fraction of the soil itself provides an adequate structure and from this point of view the content of organic matter is not of great importance. However, these soils occupy only a small area of Britain. Other soils may be more or less unstable, more or less liable to break down physically, and here organic matter is of critical importance.

Plants require soil for two main purposes. They need anchorage and they need nutrients. Anchorage is no great problem, for roots

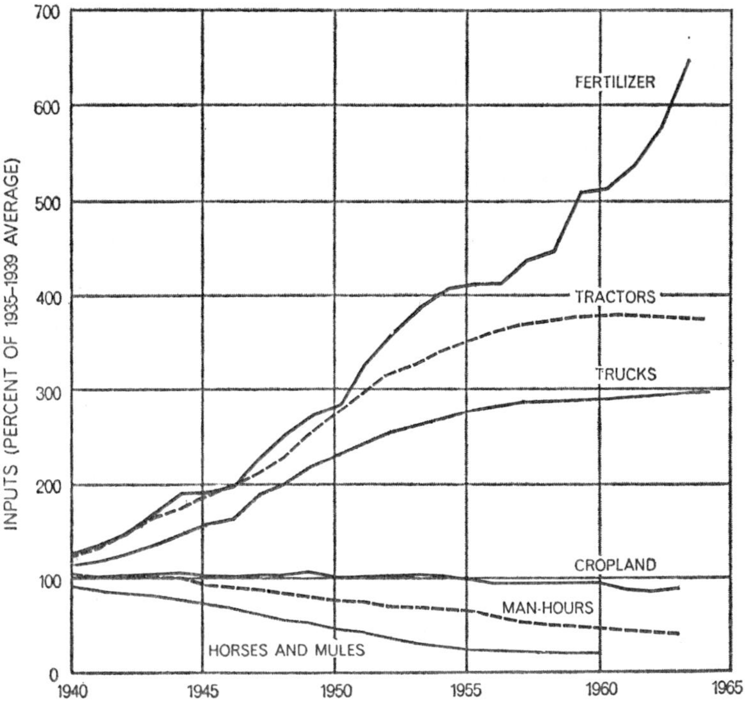

US Agricultural inputs. Current inputs are compared with those for the period 1935–9, expressed as percentages of the average for five years in each category. Increases in inputs have been accompanied by increases in output.

will penetrate quite hard layers of soil to obtain a purchase. Nutrient availability is another matter. Plants extract nutrients in a liquid form, dissolved in water. Thus it is essential that water flow fairly freely through the soil so that nutrient salts may circulate. The soil must be porous and this porosity is provided by the holes, tunnels and channels made by roots, worms and other organisms. The soil itself will tend to form particles by accretion. The size of these particles will vary according to the mineral nature of the soil, from the sandy soils in which individual grains are so large that accretion is difficult, to clay, whose minute grains accrete readily to form large lumps. These lumps, or crumbs, will become coated with microbial colonies which tend to preserve them and to prevent them from joining one with another to form a single mass. Similarly, the holes and tunnels left by animals and roots will also be coated on their insides in the same way and this will tend to preserve them. So, to some extent, the structure of most soils depends on the presence of a viable micropopulation. This, in turn, requires nourishment, which is provided by the organic wastes of plants and animals. The organisms that decompose the root are related closely to those that line the hole left by the disappearance of the root. So the organic matter content of soil becomes relevant in the water economy.

The micropopulation requires air, for the decomposition of the organic matter is an aerobic, oxidizing process. So the structure of the soil must be such that air, too, is able to circulate between the soil crumbs. In the absence of air the decomposition of organic matter may be inhibited or it may change from an aerobic to an anaerobic process. While the by-products of aerobic decomposition are plant nutrients in an available form, the by-products of anaerobic putrefaction include toxins that are harmful to many organisms, including plants.

So, on two major counts, the organic matter content of most soils is important to their structure, and their structure is important to their health.

The Agricultural Advisory Council report states:

'On unstable soils, the influence of organic matter is all important . . . Some soils are now suffering from dangerously low levels and cannot be expected to sustain the farming systems which have been imposed on them.'

There is a problem in that scientists cannot agree how much organic matter is necessary to structure. Over the years estimates have tended to go down – perhaps as the organic matter content of the soils themselves has gone down. At one time 5 per cent was considered desirable, then 3 per cent. Now there are many soils which have less than 1 per cent of organic matter, and the warning contained in the Report has been repeated on a number of occasions by the Chief Agricultural Advisor to the Ministry of Agriculture, Sir W. Emrys Jones.

While there is disagreement on the amount of organic matter soils should have, some of the effects of reducing it can be measured. The Soil Association is a private, voluntary organization that began to compare different farming systems more than twenty years ago. An area of about 200 acres was divided into three farms, each of which has been managed according to a different system; and still is. Two of the systems under consideration carried livestock, and therefore grass, the third did not. Of the two mixed farms, one has received no artificial fertilizers or crop sprays during the entire period. The soils were all similar chalky boulder clay. Today their structure differs considerably from one section to the other. In September 1970, the British Society of Soil Science published the results of its own examination of the state of the Soil Association's soils. For convenience the Association has labelled the three farms 'organic' (livestock and no agrochemicals), 'mixed' (livestock and agrochemicals) and 'stockless' (agrochemicals and no livestock).

The Society found a marked difference in organic matter content, the organic section having nearly twice as much organic matter as the stockless section and a higher carbon-nitrogen ratio. The organic soil was less dense than that of the stockless section. The Society compared the soils for their water-holding capacity and found the percentage on the organic soil was 66.2, on the mixed 55.8 and on the stockless 47.3. Other tests showed that the stockless soil was more liable to break down under physical stress; the soil was less stable.

This suggests that while the inherent stability of soils dictates the amount of organic matter they need to maintain their structure, the organic matter content also dictates their actual stability. The two are interwoven inextricably.

The differences between the organic and stockless soils can be explained by the absence of grass from the stockless rotation. More interesting are the differences between the organic and mixed sections. The mixed section soil is rather less stable than that of the organic section. Yet the only difference between the farming systems has been the use of agrochemicals on the mixed section. For many years the yield of the crops grown on the mixed section was higher, but in 1971 this situation was reversed by the introduction, for the first time, of imported organic manures (poultry manure) on to the organic section. Does this suggest a causal relationship between fertilizer treatment and stability? Some agriculturists believe firmly that such a relationship does exist.

So it seems that the soils of a large part of England and Wales are deteriorating and it appears inescapable that in time a deterioration in structure must reduce nutrient availability – regardless of the nutrient source – and so fertility. We see, too, that yields are no longer increasing. It would appear that British farming has become as intensive as it can be, that productivity has been pushed as far as possible under existing systems and techniques and that short of a major breakthrough – which might be counterproductive since it would require increased fertilizer use and increased intensification – Britain will be fortunate if its food-producing capacity does not decline in the years to come and, probably, stabilize at a level somewhat lower than it is today. At present, in spite of all its achievements, Britain produces about half of the temperate foodstuffs it consumes. By the latter part of the century it will be able to feed less than one third of the population that has been forecast.

What are the implications of this for developing countries? Britain, as we have seen, enjoys almost ideal farming conditions. The climate is mild, rainfall is adequate and distributed fairly evenly through the seasons, there is ample sunlight, yet increased production by intensification has reached, and probably passed, its limit. If the climate were more severe, the temperature higher, the days shorter (which they are in summer in the tropics), the rainfall so distributed that a dry season were followed by a wet season, might not the limit of productivity be lower? If Britain had not experienced the centuries of good farming that created large reserves of organic matter in its soils, might not its present losses of structure have occurred sooner and at a lower level of intensification? If British soils are beginning to erode, which it seems they are, either by drying out into something approaching dustbowls – there are dust storms every year now in parts of the East Midlands and East Anglia – or by waterlogging into something approaching marsh, might not these effects have appeared sooner on less stable soils? Is it not probable that the intensification of production through an increase in the inputs of chemical nutrients might lead to the depletion of good soils, rendering them less stable, and the accelerated erosion of poor soils, where erosion is a problem already?

Yet even with all the many resources, both natural and technological, that have been available to them, British farmers are still able to feed only half the country's population.

All of this would suggest that if an intensified farming system on anything remotely approaching the scale of intensification that has been imposed in Britain were to be attempted in much of the third world, it could not hope to succeed.

Unfortunately, the implications are more serious even than this. If increasing intensification tends to deplete soils, then it is subject to diminishing returns. If farmers are encouraged to intensify, they may produce increased yields for a limited period, but in doing so they will prejudice their chances of maintaining even the yields they produced before. As the soil deteriorates, yields will decline. As yields decline the farmer's income will decline also. As the farmer grows poorer he will be less able to afford the chemical inputs on which his crops now depend, because he has been extracting nutrients more quickly than they were accumulating and being made available naturally in the soil. His yields will decline still further. He will be forced to abandon the new varieties and return to the old ones, but on his now relatively impoverished soil, he will be producing less than ever. Meanwhile, attempts to intensify on eroding or unstable soils will have aggravated existing problems and some land will have been lost to agriculture altogether. So, while the population continues to grow – and to occupy agricultural land for urban development – food needs will have to be met from dwindling amounts of deteriorating land. There is a grave danger that in the long run – if there is a long run – intensification may result in a situation worse than that which obtained before it began.

8

Take care, O dead Earth

The New Earth Villagers, the young people from all over the world who attended the FAO's Second World Food Congress in The Hague in June 1970, at the invitation of FAO, distributed a sheet, bearing a crucifix and the following words:

'Nature is *our* capital. Hold the interest of it. *Take care!* O dead earth.'

The pollution and general degradation of the human environment is, decidedly, a live issue. It is also a new one. It is only since the mid-1960s, or thereabouts, that the environment and its health have become fashionable areas of concern. 1970 was European Conservation Year, sponsored by the Council of Europe, and the interest that had been building up steadily over several years was given a further impetus. Now scarcely a day goes by without some news story, some article, some broadcast, that describes or comments on pollution. Songs have been written about it, comedians make jokes about it, politicians make speeches about it. It is 'in' and it is safe, for no one can criticize those who express concern about the quality of life of their fellow citizens. Several countries, including Britain, have a Ministry of the Environment, and most have an environmental research department with official status. Why all this fuss?

There are some who dismiss it all as a fad, who argue that man has always polluted his environment and always will. Nothing has happened so far to do him great harm, so why should we believe anything ever will? Certainly, there are areas of land that have been made unsightly by industry. They are aesthetically displeasing, but this was a necessary price to pay for economic advancement and the benefit can be measured in terms of raised standards of living for working people. In any case, that was all long ago. We put a price on environmental quality nowadays and it presents no great technological problem to build factories that are pleasing to the eye, that blend into the natural landscape, if that is what is required. It is possible to clean the effluent discharged by industry. Perhaps we will have to pay a little more for industrial products, but we can afford that. Deep down, all is well. Is it?

On 14 April 1970 – European Conservation Year – people living close to a part of the Manchester Ship Canal had telephoned the police complaining of a smell of gas. At about 7.15 in the morning the police were calling at various houses to check the complaints. One policeman was standing by the bank of the canal. A mist hung over the water and through it he could see a small boat in mid-stream. There was a group of men in the boat and they seemed to be in trouble. One looked as though he was unconscious. They were having difficulty steering and the boat was drifting aimlessly. As he watched, the policeman saw another man climb into a boat and begin to row out from the bank, apparently to their aid. The man's name was Bernard Carroll.

Bernard Carroll was twenty-four years old and he worked as a ferryman. He never reached the boat with the party of men. When he was some 20 or 30 yards from it *the water* burst into flames. The fire spread rapidly along 1,600 yards of the canal. It engulfed both boats. Five men survived and three were never found. Bernard Carroll's body was recovered. The inquest said he had been burned to death. No one knows how the fire started.

Every day the Manchester Ship Canal receives all the polluted water from the rivers that flow into it, and, in addition, 90 million gallons of sewage and 180 million gallons of industrial effluent.

It is possible that the water can become so heavily polluted that gases bubbling up through it may ignite. If this is what has happened to the canal, there is a precedent. The River Cuyahoga, which flows through Cleveland, Ohio, is the first area of water in the world to be classed officially as a fire hazard. It has burst into flames on a number of occasions. Once it nearly destroyed two of the town's bridges that cross it.

So this is what pollution is about. It is about people dying. When I was a child I was encouraged to play out of doors whenever the weather was fine. At school I was turned out into the playground at break times and for the whole of the lunch period that was not spent eating. Today, in many cities throughout the developed countries, warnings are broadcast when the air pollution reaches a certain level. Then children are not allowed out to play. It is dangerous to play outside. It is dangerous to breathe deeply. This is what pollution is about.

At the 1971 annual conference of the British Association for the Advancement of Science, held in September in Swansea, it was reported that University College, Swansea, was working on a new early warning system for the detection of certain substances in the air that are highly toxic to man. Particles of lead, cadmium, nickel, zinc, copper, cobalt and other materials have been identified in air, soil, vegetation and wildlife. The development of this system is being treated as a matter of urgency. The Medical Research Council has called a meeting to examine evidence of the extent of lead pollution in the atmosphere and its effect on health.

In an interview published in a magazine called *Industrial Management*, the Duke of Edinburgh said:

'I foresee the time when there will have to be a code of practice accepted by everybody. It will have to be backed by legislation and given some sort of sanction because everybody's hands need to be tied, not just industry's.'

The Duke went on to explain that, in his view, pollution was not caused by, nor was it the sole responsibility of, industry. It is linked, he said, to population growth and to the pressure exerted on the environment by population.

Obviously, this is true, but, as we have seen, population growth and industrial growth cannot be wholly separated. It is no coincidence that the period of greatest pollution has occurred at the same time as the period of most extensive industrial growth. It is unlikely, therefore, that pollution can be controlled finally and to a satisfactory degree, without curtailing industrial growth and the growth of population. There is an economic reason for this. The cost of preventing the release into the environment of pollutants relates directly to the degree of cleanliness that is required. The cost rises very steeply as the last few per cent is controlled. The effect on the environment, however, is not that of a single factory, or a single industry, but of all contributors taken together. Therefore, if the level of pollutants is to be kept to a satisfactorily low level, the control exerted by each factory will relate to the number of factories.

The more there are, the cleaner each of them must be. Since pollution cannot be prevented entirely no matter what controls are instituted, when there are enough factories, pollution will exceed the desired level. Meanwhile, as the number of factories increases, the cost to each of them of controlling the pollutants it produces will rise to a point at which it becomes economically impossible to contain within the industrial operation. The only solution, therefore, is to restrict industrial growth, with all that that implies. Is it worth it?

The situation is grave. There has been high level scientific discussion of the possibility of our triggering major climatic changes through pollution. The quantity of fine particulate matter released into the upper air has been increasing since industrialization began. Now it has formed a blanket that covers the northern hemisphere. The fine particles, too light to fall, reflect back into space incoming solar radiation. Should it increase still further – and it is increasing all the time – and if large quantities of water vapour are introduced into the dry upper air as they would be, for example, were large fleets of high altitude supersonic aircraft to go into commercial service, the water vapour might condense on the particles – this is how clouds form. The clouds would form thin, icy layers similar to cirrus or cirro-stratus and they would increase the amount of incoming radiation that is reflected back into space. Mean surface temperatures would fall. By how much would they need to fall before the sea temperature in the arctic and antarctic became low enough to cause the icecaps to grow? Were this to happen, the process would be likely to trigger a chain reaction and we would face an ice age. Alternatively, the combustion of fossil fuels releases carbon dioxide into the atmosphere. Levels of atmospheric carbon dioxide are increasing. Carbon dioxide permits incoming solar radiation to pass through it, but it traps radiation reflected back from the earth's surface, so raising the air and then the surface temperature. This is known as the 'greenhouse' effect. By how much does the temperature need to rise before the icecaps begin to melt? Were this to happen, again it might trigger a chain reaction and with the consequent reduction in the solar radiation reflected back into space by such large areas of ice and snow, the climate would soon become warmer. Areas that are cold might become temperate; areas that are temperate might become hot – and dry – and areas that are hot and dry already would become still hotter and, if this is possible, still drier. Meanwhile, the raising of the levels of the seas would inundate many of our cities, as we saw when we considered the possibility of melting the icecaps deliberately to increase our supply of fresh water.

The seriousness of the problem is evidenced by the fact that such possibilities as these can be, and are, discussed.

This is pollution. The dangers with which it threatens us are serious enough to call into question the continuation of industrial expansion at anything like the present rate, even were the resources available to permit it.

The production of pollutants is only one part of the problem;

CO_2 produced by Fossil Fuel combustion 1950–67 (thousand million metric tons)

YEAR	Coal	Lignite	Refined oil fuels	Natural gas	TOTAL
1950	3.7	0.9	1.4	0.4	6.4
1951	3.8	0.9	1.7	0.5	6.9
1952	3.8	0.9	1.8	0.5	7.0
1953	3.8	0.9	1.9	0.5	7.1
1954	3.8	0.9	2.0	0.6	7.3
1955	4.1	1.0	2.2	0.6	7.9
1956	4.4	1.1	2.4	0.7	8.6
1957	4.5	1.3	2.5	0.7	9.0
1958	4.6	1.4	2.6	0.8	9.4
1959	4.8	1.4	2.8	0.9	9.9
1960	5.0	1.4	3.1	1.0	10.5
1961	4.5	1.5	3.3	1.0	10.3
1962	4.6	1.5	3.5	1.1	10.7
1963	4.8	1.6	3.8	1.2	11.4
1964	5.0	1.7	4.2	1.3	12.2
1965	5.0	1.7	4.5	1.5	12.7
1966	5.1	1.7	4.8	1.6	13.2
1967	4.8	1.7	5.2	1.7	13.4
1980 (estimate)	11.1		10.8	4.0	26.0

(*A Blueprint for Survival*)

Estimated concentration factors in Aquatic organisms

radionuclide	Phyto-plankton	Filamentous algae	Insect larvae	Fish
SITE—*Columbia River*				
Na^{24}	500	500	100	100
Cu^{64}	2,000	500	500	50
Rare earths	1,000	500	200	100
Fe^{59}	200,000	100,000	100,000	10,000
P^{32}	200,000	100,000	100,000	100,000
White Oak Lake				
P^{32}	150,000	850,000	100,000	30–70,000
Sr^{90-90}	75,000	500,000	100,000	20–30,000

(*A Blueprint for Survival*)

World Average annual rates of increase for the period 1951–66 for selected aspects of Human activity

	Percentage
Agricultural production	3
Industry based on farm products	6
Mineral production (including fuel)	5
Industry based on mineral products	9
Construction and transportation	6
Commerce	5

(UN Statistical Yearbook, 1967)

they must be distributed in the environment. Smoke released from a factory chimney, polluted water or sewage entering a river, will be distributed by the natural flow of the fluid in which they are carried. In other cases the consumer of the product may effect the distribution himself. All the pollution caused by motor vehicles is distributed by the users of those vehicles. Litter and garbage is distributed by people.

Nevertheless, all pollution of this kind can be regarded as an undesirable side-effect of a process that itself is desirable. What if the process itself includes the distribution of pollutants? There is one such industry.

Agricultural chemicals are biologically potent – that is why they are used. They stimulate growth, or they over-stimulate it, or they retard it, or they kill. They are used to control by force natural processes. This is their sole purpose and their only use. If pollution of the environment it defined as the introduction into the environment of materials that interfere with life processes – and pollution can be defined in this way – then modern agriculture is a polluting industry by definition. This is not to condemn farmers out of hand, or to demand that they abandon their present practices forthwith, but it is to remind us that they manage very large areas of land – 80 per cent of Britain's land area is devoted to agriculture or forestry in one form or another – and because of this and because of the nature of the substances they employ, they have the potential to cause more pollution than any other section of the community.

In June 1969, a quantity of an insecticide, endosulfan, was released into the River Rhine. Millions of fish died and several major Dutch cities turned to emergency water supplies as the dangerously polluted water flowed through Holland into the sea. Endosulfan is a persistent organochlorine insecticide and a spokesman for the company that manufactures it stated at the time that as little as 100kg could have caused the Rhine disaster. It could have been poured into the river by a careless farmer. It could have resulted from an accidental seepage from the factory, which is situated on a tributary close to where it joins the Rhine. It could have been spilled when a barge was involved in a collision. There are a hundred ways in which it could enter the river. The Germans and the Dutch are dependent on the Rhine for water. Fortunately they were able to turn to alternative sources of supply. They are not dependent for food on the fish that were killed. Had Germany and Holland been developing countries, it is almost certain that those fish would have represented an important food supply.

Suppliers of pesticides supply, along with their instructions for use, warnings and safety instructions. The College of Agriculture of the University of Nebraska has produced a small booklet on the safe use of insecticides. It advises: 'study the label; it will tell how much to use, what crops to use it on, waiting periods following use, and safety precautions that must be followed. Keep insecticides under lock and key. Do not breathe dusts or sprays; mix and apply insecticides so that dust or fumes are blown down-wind. Be certain drift does not reach other crops, livestock, or inhabited areas. Do not spill sprays or dust. Always keep insecticides in original containers. Wash and change clothing; after applying insecticides, wash thoroughly with soap and water, and change to clean clothing. Launder clothing after each day's use. Never wear contaminated clothing. Always wash before eating or smoking. Check for leaks. Never combine insecticides. Keep a record; always keep a record of material and amounts used, and date of application. Avoid drift. Destroy the container.' Then follow instructions in the event of accidental poisoning – instructions for those first on the scene and for the physician.

The cheapest and most widely used insecticide is DDT. At the present time it is on trial all over the world. In the developed countries, as more and more is learned of its side- and long-term effects, more and more controls are imposed on its use. Yet in the developing countries it is often the only insecticide that is available. According to the FAO, the developed countries cannot afford to abandon DDT, although it urges that its use be kept to a minimum. In the view of the FAO, there would be a disastrous reduction in crop yields if it were to be abandoned too hastily. Dr. Norman Borlaug has criticized those who would have DDT withdrawn because of the disastrous effect this would have on food production. The WHO, on the other hand, calls for reductions in its use and has called it 'a scourge of the natural environment.' Yet even the WHO does not advocate a complete ban, because of the increase in the incidence of malaria it believes would follow – and which did follow when it was banned in Ceylon.

DDT is persistent. It has a half-life of ten years, which means that half of any quantity released into the environment will have broken down and disappeared ten years after application. Even then, the harmlessness of the breakdown products is in some doubt. They include DDD and DDE, both of which are themselves organochlorine insecticides. The persistence of DDT has led to its dissemination all over the planet. Traces of it have been found in Antarctica and in Greenland, far from any site of application. At the present time it is estimated that roughly 100,000 tons of DDT are produced each year – one quarter of the world's total pesticide output.

Studies of the residual effects and the build-up of DDT have shown that in fish, which concentrate residues in their bodies, levels of concentration do not reach a peak until 11 years after application. DDT was introduced in 1940 and usage has increased steadily since, so that maximum annual use to date is in 1971. If DDT were banned all over the world, and no more was used after the end of 1971, the concentration in fish would not reach its maximum until 1982 and it would be 1996 before it fell below its 1971 level.

Apart from residues in the air following spraying and as a result of evaporation from the ground surface and residues in soils and in food, most of the residual DDT enters the environment via water as it is washed out of soils. In common with the other organochlorines it is insoluble in water, but highly soluble in oils and fats. Since it does not deteriorate, and since it cannot be metabolized, it tends to dissolve in fatty substances that are incorporated into animal tissue. There it remains for some time. As small animals are eaten by larger animals, the DDT content of each of them is ingested by the larger animal and accumulates in its body. So, as we move up a food chain, DDT concentrations increase, animals at each level retaining the DDT from all the smaller creatures they have consumed. Fish are at the end of the first food chain in water and so they are the first animals to be exposed to DDT as it enters the environment by this route. Man, too, is at the end of a food chain. The delay in the emergence of peak concentrations is due to the time taken by the DDT to find its way into the water from the soils to which it was applied, and to the much shorter length of time required for its accumulation up the food chain. The implication of these studies is that if severe environmental damage were to be associated with DDT, so that its use were prohibited, high levels of DDT contamination would remain for 30 years.

The case against the use of DDT on the grounds of its effect on wildlife is based on a mass of evidence. Together with the other organochlorines, it has reduced the fertility of birds of prey and it has caused sharp reductions in their populations. This is of more importance than it might seem, for it is the predators, those at the end of food chains, that control the populations of all those below them. When the predators leave there may be ecological disruption whose results are incalculable: an increase in pest species is one obvious possibility. Fish are particularly susceptible to it and have been poisoned directly. More sinister is its effect on planktonic algae in the sea. Concentrations as low as one part per billion (million million) prevent photosynthesis in these minute organisms. This may have an adverse effect on fisheries, because it is the plankton that are the base of marine animal life, and it may possibly have an effect on the world's balance of atmospheric oxygen, because although it is not known with accuracy, the contribution of marine plants to the oxygen we breathe – all of which is replaced by plants – may be considerable. After all, the seas cover 71 per cent of the surface of the planet. Professor Jacques Piccard, the Swiss underseas explorer, maintains that two thirds of the world's atmospheric oxygen is derived from the sea, and in his view oceanic pollution that killed the plankton could suffocate life on earth.

It is understandable, therefore, that the introduction of the full range of modern pesticides, and an increase in the use of DDT, in developing countries, should be a cause for alarm among biologists.

On 12 October 1970, the FAO announced plans to establish a Pesticide Research Centre for the Developing Countries. The Centre will cost 2 million dollars and it will investigate 'hazards to users, consumers and the environment of pesticides of actual or potential interest to developing countries because of their low price and low toxicity.' The FAO's Committee of Experts on Pesticides in Agriculture, 'expressed concern on the very general practice of using pesticides as "insurance" against any likely future pest or disease which may never materialize. This tended to inflate costs and pollute the environment. Applications of pesticides should be done according to "good agricultural practice" only when pests were in sufficient number to warrant it and never when a chemical was likely to persist long after harvest time.'

Pesticide manufacturers have sponsored conferences and seminars on the safe and effective use of pesticides. One was held in Sao Paulo, Brazil, in May 1971, chaired by Mr. J. I. Hendrie, Head of the Agricultural Division of Shell, with the co-operation of Bayer, Ciba, Geigy, Dow, Hoechst, ICI, Phillips, Shell, Hudson, Stauffer and Sumimoto, all major producers. It seems reasonable to suppose that the real aim of such seminars is to promote the use of pesticides in developing countries and this is, of course, the policy of the FAO.

We have considered already the likely long-term contribution to food supplies that pesticides can make, and we have considered, too, the likelihood of their being introduced in the developing countries on anything approaching the scale at which they are used in the developed countries. Nevertheless, if they were to be introduced, what would be the effect on the environment of the areas using them, and of the world as a whole? It is said, by the manufacturers, that it is the mis-use, rather than the use, of pesticides that is harmful, and that with care and discrimination they can be used safely. Even supposing this were true, is it reasonable to suppose that farmers in developing countries would use them more discriminatingly than those in the developed countries? Already there have been cases of severe poisoning as a result of pesticides being handled and distributed by workers who were not able to read the labels. In Britain children have died from drinking paraquat, a 'safe' herbicide that had been transferred from its original containers into, of all things, lemonade bottles. Is there a difference? These are instances of acute poisoning, but evidence is beginning to emerge to suggest that even the smallest doses, repeated over a prolonged period, can have an adverse effect on human health. A report published by the US Department of Health, Education and Welfare back in 1969 suggested that aldrin, dieldrin and DDT may be carcinogenic.* It appears that organochlorines including DDT interfere with hormonal balances in the body and that they affect the metabolization of certain drugs used in medicine. It is also possible that they may affect the central nervous system, producing symptoms identical with those of a whole range of nervous disorders.

All the indications are that the adverse environmental effects associated with pesticides are related directly to the amounts used and that if their use increases these effects will increase in proportion. Thus, any increase in pesticide use is harmful and if they were to be brought into use in the third world at rates and frequencies of application similar to those considered necessary in the developed

* *Report of the Secretary's Commission on Pesticides and Their Relationship to Environmental Health.* H.E.W., 1969, p. 471.

countries, even the most stringent precautions could not prevent gross, perhaps catastrophic damage.

It is not only with pesticides that farmers harm the environment. Fertilizers, too, can be dangerous.

Fertilizers are useless without water to dissolve them and so to make them available to plant roots. The capacity of the soil to retain water, with or without fertilizers, is determined by its physical structure and on many soils this, in turn, relates to the quantity of organic matter present. If the organic matter content is reduced, so is the water-holding capacity, and so is the efficiency of fertilizer. The reduction in the content of organic matter and the consequent loss of soil structure is an inevitable result of arable farming without livestock and its extent is related directly to the intensity of the farming system. A reduction in the efficacy of fertilizer usually leads to an increase in the rate of application, to ensure that a constant quantity reaches the plant. The balance that is not taken up by the plants still dissolves, however, and as the water that is not being retained by the soil runs off into the nearest waterway it carries the fertilizer with it. Estimates vary as to how much fertilizer is lost in this way. It has been stated at a symposium on Potential Crop Production held at the Plant Breeding Station in Aberystwyth, that beyond a certain optimum rate of application – which has been reached or exceeded by many British farmers – all additional fertilizer ends up in rivers or lakes. The Centre for the Biology of Natural Systems at Washington University, St. Louis, Missouri, has found that in one area it investigated 50 per cent of the nitrogen fertilizer applied was taken up by the plants, 25 per cent remained in the soil, and the remainder was either leached into water or transported into the air.

The first and most widely described effect of fertilizers in water is that of eutrophication. Literally, this means 'over-enrichment.' Strictly speaking, it is not pollution at all. The pollution is a secondary effect.

In water, predictably, plant nutrients stimulate the growth of aquatic plants. These may be weeds that are there already, or they may be unicellular algae. Very large numbers of algae, which have a very short life cycle, can bloom, literally, overnight. As the plants die, they are decomposed by bacteria present in the waters, whose numbers increase to respond to the increase in their food supply. The bacteria consume free oxygen and the water becomes de-oxygenated. Fish, which depend on very small quantities of free oxygen, die from asphyxiation, adding to the decaying organic matter. As the process continues, all animal life is extinguished. When the availability of free oxygen falls below the level necessary for the survival of the bacteria, aerobic decomposition is no longer possible and the process becomes anaerobic – putrefaction. Putrefaction releases toxic by-products into the water and so more living organisms are killed.

The final stage of this chain of events may be reached when the water will sustain life at only the most rudimentary level. Parts of lake Erie, the world's most notorious example of eutrophication, sustain little life beyond the algae, which continue to bloom so that at times the water has the colour and consistency of a thick pea soup.

In some still lakes the water may become stratified. During the summer, the upper layers are warmed by the sun and the lower layers remain cooler. The algal blooms at the surface are replaced by fresh growths and sink slowly to the cooler layers below, where they begin to putrefy. In autumn the movement of colder air across the surface cools the upper layers of the water and a point may be reached at which the upper layers are cooler than those below. The denser, cold water will sink, the less dense, warmer water will rise, and the whole lake will turn over. The putrefying organic matter held in the lower layers is brought to the surface.

If the water is shallow, the accretion of organic matter on the bed, combined with increased water loss from transpiration by the plants growing in it may cause the water to dry up altogether.

Eutrophication is an ageing process. All areas of still, or relatively still water are subject to it. All fresh water contains some plant nutrients. Over long periods of time these will accumulate. However, the natural death of a large lake from 'old age' is a process measured in biological time – thousands of years. By adding large quantities of plant nutrients we have succeeded in speeding up the whole cycle.

Of course, fertilizers leached from farm land are only one source of plant nutrient. Sewage is another major source and the phosphates contained in detergents are a third.

Direct eutrophication is only one hazard caused by the addition of nitrogen to water, however. We have considered the uptake of excess nitrates by plants which can lead to methaemoglobinaemia in infants. Nitrates can be held in water for long enough for them to reach human consumers. There have been instances of babies dying from methaemoglobinaemia caused by the ingestion of excessive nitrate from domestic water supplies. At times this danger has been serious enough in some areas of the US for bottled water to be recommended for infants. It is considered that a nitrate content in excess of 50 parts per million (ppm) in water may be harmful and in Britain a limit of 30ppm is set. In parts of lowland England which are devoted to intensive farming and where river water is abstracted for drinking purposes, such as Suffolk, the nitrate levels have exceeded 20ppm at times and it has been suggested that limits may have to be imposed on nitrate fertilizer application rates in these areas in order to safeguard the water supply, even if this restricts agricultural productivity. Even then, the British limit may be on the high side. Leo Michl, of the Illinois Public Health Department, told a meeting organized by the Center for the Biology of Natural Systems in November 1970, that 'if you want to protect all the people, you can't put any added inorganic nitrogen in the water . . . if you want to protect most of the people, the PHS (Public Health Service) standard of 10ppm nitrogen is safe.'

It is only in acute, or clinical, cases that methaemoglobinaemia causes death. There is growing evidence that indicates that at sub-clinical levels this condition reduces the oxygen available to tissues and that this affects certain bodily functions. In particular, it causes brain damage in foetuses. Professor Barry Commoner, of Washington University, St. Louis, told the same meeting that imbalances caused by nitrogen in water can lead to the exposure of populations to a whole range of environmental diseases such as meningoencephalitis, caused by a common soil amoeba which, because of exploding algal growth in water, moves into the water where man may come in contact with it.

'Using the example of bubonic plague, which is presently carried by fleas on the back of rats in certain Nevada caves,' said the official report of the meeting, 'Dr. Commoner pointed out that one of the things which prevents an epidemic is the isolation of these caves. In the case of the nutrient enrichment of surface waters, what one finds is a breakdown in the isolation between land and water, a drastic alteration in the environmental system which transforms the barrier into a pathway by which the disease can affect man.'

It may be fanciful to think of bubonic plague as a possible side effect of farming in Britain, but it is less fanciful, perhaps, in tropical developing countries. We should remember, too, that the leaching of nitrogen and other plant nutrients into water is a serious problem in this country, where soil and climatic conditions are ideal for its uptake by plants. If we imagine a soil less able to retain moisture, a soil from which water runs off quickly, combined with a large seasonal variation in rainfall, then we may expect the problems to be correspondingly greater if fertilizer use should increase to anything like European or American levels. If the IWP target of 90lb per crop per acre were achieved, the effect on water might be disastrous. At a plenary session on environmental pollution that was held during the FAO's Second World Food Congress, Lord Ritchie-Calder warned of what might happen were fertilizers to be used on a large scale in parts of India from which they could be washed by the monsoons into the Bay of Bengal. It would produce a complete change in the ecology of the Bay. It is only in the most fertile areas, those with the best soils, that two or three crops a year could be grown, but in those areas the contamination of some food crops, the waste of precious fresh water through increased plant growth and subsequent transpiration, and the contamination of river and lake water used for domestic purposes and as a source of fish protein might begin by outweighing the advantage gained by increased crop yields and might end by triggering ecological reactions on a scale too vast and too complex to be predicted or controlled.

The FAO is aware of this danger and the 1972 Stockholm Conference will devote time to representatives of the world's fertilizer manufacturers who will take part in discussions aimed at determining the extent of existing pollution from fertilizers. The discussions will be sponsored by the Fertilizer Industry Advisory Committee of the FAO.

At the present time agricultural pollution in the developed countries probably has little or no effect on the countries of the third world except to the extent to which it pollutes the oceans. It is here that it joins forces with urban and industrial pollution.

Dr. Bruce W. Halstead, of the International Biotoxicological Centre, Colton, California, and a WHO consultant, told a meeting of the UN Interagency Advisory Group of scientists on marine pollution in March 1971, that 'toxic pollutants enter living systems through biological concentration. The phytoplankton acts as a biological blotter picking up nutrients, trace metals and toxic pollutants. Organisms feed on the phytoplankton and successively pass on the pollutants to higher organisms. As this process continues up the food chain, concentrations of an intoxicant reach their highest levels in predators, such as marine mammals, birds and man.' Dr. Halstead said that 1,000lb of phytoplankton were needed to produce 100lb of shellfish, 50lb of small fish, 10lb of small carnivores and one pound of the carnivorous fish most commonly harvested by man. 'When man eats a fish he is receiving the combined toxic chemical concentrates at the end of the food chain. These chemical pathways proceed in many directions throughout the food web and along these avenues toxic substances flow from one organism to the next.' The build-up of toxic substances in the marine environment is occurring with great rapidity owing to population and industrial growth and Dr. Halstead said that 'prompt and drastic steps' were needed to safeguard the seas.

Thor Heyerdahl, the Norwegian explorer, who sailed his papyrus raft the Ra across the South Atlantic, reported severe pollution far away from the nearest coast and a long way from the main shipping lanes. His crew had to abandon their practice of dipping their toothbrushes in the sea because of huge rafts of oil and debris floating on the surface. Some of this matter had been discharged at sea, but a large part of it had been carried down river systems from industrial regions.

Already, some seas are suffering from gross pollution. The Baltic, the North Sea and the Mediterranean receive far more wastes than they can absorb, break down or disperse. Man has always used rivers and the seas as sinks into which he has thrown his garbage. Now there are too many of us, and we produce too much garbage. The Swedes have been warned by their government of the dangers of eating fish caught in the Baltic because of the high levels of mercury they contain. Americans are worried about mercury in fish caught in the Pacific and about cadmium residues in fish from the Hudson River, New York and from the Atlantic coastal waters. Cadmium can cause anaemia, high blood pressure, gastritis, kidney and nerve damage. Pesticides, particularly DDT, are killing fish and one by one the industrial nations are coming to accept the need for international agreement to prohibit the dumping of wastes in the sea.

The Times carried a report on 12 June 1971, describing a new form of pollution:

'Even in the unpolluted Pacific area near Hawaii large amounts of carbon monoxide have been found in rainwater. This points to a natural process manufacturing the poisonous gas inside clouds, and adds further to the concern over man's large-scale pollution, which must be viewed with greater alarm coming on top of large-scale natural pollution.'

It was known already that the Atlantic Ocean contains dissolved carbon monoxide and scientists had estimated that about 5 per cent of atmospheric carbon monoxide pollution is natural in origin. The rainwater samples contained up to 200 times more dissolved carbon monoxide than could be explained by natural absorption. The actual content of the samples averaged 0.14ppm.

In the industrial regions of the world, rainwater has been polluted for years and Sweden has complained about the acidity of its rain caused, apparently, by polluted air from Britain.

All the developed countries are worried about the pollution of their environments. Their reactions include speeches, conferences, promises and even resentment. Dr. A. Michaelis, science correspondent of the *Daily Telegraph,* has said that scientists, particularly chemists, need to protect themselves by a public relations exercise designed to give them what he called 'an anti-pollution image.' This could be achieved, he said, if they devoted – and were seen to be devoting – more of their time to devising solutions to pollution problems.

The country that has advanced farthest in the field of environmental control is Sweden, where pollution is regarded as one part of a much wider environmental protection problem. Measures to deal with it are co-ordinated at government level and legislation has been introduced covering the discharge of sulphur dioxide from oil-fired heating systems, motor exhaust fumes, noise, pesticide use and seed dressings. The legislation is backed up by an elaborate administration headed by the Ministry of Agriculture, the Ministry of Communications and the Ministry of Health and Social Affairs. Apart from legislative control, Sweden has adopted the principal that the production or use of substances detrimental to the environment should be subject to a tax levy aimed at adjusting prices to take into account the social and environmental costs borne by the community at large. The effect is to bring about a shift in demand to less harmful products and to encourage more favourable methods of production. The whole programme is accompanied by government-sponsored radio and television programmes that explain the need to protect the environment.

Sweden is a large country with a small, relatively wealthy population. Its experience and expertise can provide invaluable guidance for other countries with pollution and environmental problems, but it may be that Sweden is exceptional in its ability to resolve them.

Pollution is a product of people and of industry and for so long as nations accept uncontrolled population growth and aim for industrial expansion and associated higher levels of consumption, it is difficult to see how the environment can be protected.

What of the developing countries? They are aware of the environmentalist movement, but there is a grave danger that they may see it only as a device whereby wealthy nations improve still further their living standards; as a luxury they themselves cannot afford. If they aspire to the life of the industrial countries they will have to pay the price in full. If they were to industrialize fully, however, the price would almost certainly be the total breakdown of the biosphere and possibly the end of all the higher forms of animal life – including man. It is ironic that their efforts to improve their own food supply should be attended by the very real risk of catastrophes wider, more severe and more uncontrollable than those of famine and disease.

As with most of the apocalyptic crises facing man, it is not inevitable. There is an alternative, provided limits are placed on expansionism. China has provided examples of ways in which pollution problems can be tackled, by recycling waste products and by not engaging in activities that are environmentally disruptive. Of course, the concept of recycling is not new, but in most countries it is regarded as uneconomic. The Chinese, on the other hand, argue that any materials recovered from waste will reduce pollution, protect the health of their citizens, and so represent very real gains. There have been a number of cases in which factories have reduced pollution, sometimes to zero levels, by recycling their wastes, simply because of the threat to the health of those living in the vicinity. Certainly the approach is correct and sane and, according to reports of visitors returning from China, it works.

The conclusion must be that the implementation of the IWP plans for increased agrochemical inputs in the developing countries is fraught with danger and that the road to industrialization on the pattern of the existing industrial nations would mean ecological disaster on an unprecedented scale. As it is, it is unlikely in the extreme that the developed countries will be able to reduce the pollution of their environments by industry and agriculture unless they adopt and then encourage other countries to adopt, policies that aim for stability and that reject permanent and unbridled growth.

9

Broiler houses to feed the World?

The deterioration in the structure of some of the most fertile of British soils is due to a reduction in the amount of organic matter they contain. The remedy is simple: a return to grassland and mixed farming. But why has it happened?

The cause has not been the intensification of farming in itself. There is good reason to believe that the heavy use of artificial fertilizers contributes to soil deterioration, as the use of heavy machinery does, but at most this is only a part of the story. The main reason is the conversion of large areas from mixed to purely arable farming. It is a general rule of agriculture that mixed, or stock farming improves soil structure, arable farming depletes it. As we have seen, during the few years when something like a million acres of grass were ploughed up, around 900,000 acres were added to the total area sown to barley.

The grass has been replaced by barley because the animals have left the land. In an approach to the problem of how to increase food production from animal sources, the argument ran something like this. Livestock live most of their lives out of doors and the greater part of their food consists of the grass and herbs that they graze. They can be made more productive either by increasing the number of animals on a given area of land, or by adapting them or their food so that a higher proportion of it is converted into meat, or other desired products. However, there is a clear limit to the density with which the animals can be stocked out of doors. Too many and they will so churn up the soil, especially when it is wet, that its structure will be destroyed. That situation will be reached only if the productivity of the pasture they are grazing can be increased to provide them with food. This is possible by the use of fertilizers – at any rate to some extent – but too much fertilizer and the grass becomes unpalatable. Animals will prefer not to eat it and if they should eat it they may become ill. So, try as we may, productivity will be limited so long as we stick to animals grazing out of doors on grass. Can we, then, provide a food more nutritious than grass? Yes, we can. There are cereals – which are selected and highly bred grasses – that have a higher protein content. So let us plough up the grass and grow more cereals. If we do this, however, we must find somewhere else to keep the animals, for cereals cannot be grazed in the way that grass can. The answer to that is to keep them indoors and to bring their food to them. In its most extreme form, in which the animals never leave their buildings at all, the version of this system applied to cattle is called 'zero grazing.'

There are other advantages to be derived from keeping animals indoors. While they are outside they move about. This consumes energy that could otherwise be employed in converting feedstuffs into product. There is no economic advantage to anyone in having animals moving around. Out of doors the animals must also contend with the weather. Sometimes it is cold, sometimes it is hot. Being warm-blooded creatures, they needs must maintain the homeostasis of their bodies and this, too, consumes energy. If they go to live indoors they can be kept still and warm so that energy will be conserved. So, simply by keeping an animal indoors and bringing its food to it we will increase the efficiency with which it converts feedstuffs into products we can eat – and sell.

Like other changes that have taken place in farming in the industrial countries, this development is labour-saving, and with agricultural labour in short supply this is an important factor. Less manpower is required to manage animals if they remain indoors in one place, and since so many of their physical requirements can be provided by automated systems, there is less need for skilled labour.

Having agreed, then, that it is more efficient, and therefore desirable, to keep animals indoors, where shall the units be located? The obvious answer is to site them as close as possible to the markets they will supply, but there is a further economic factor that came into play here. As farmers were squeezed to grow more and more food, they were compelled to engage in those enterprises which brought them the largest returns. This was not, generally, a matter of greed; it was one of survival in a harsh economic climate. Cereals were more profitable than livestock husbandry – particularly with regard to cattle and poultry – and so land that could grow cereals had to be used for this purpose. The livestock had to be taken to the less profitable land where, because of the great increased density with which it could be concentrated, it once more became economically possible. Generally speaking, in Britain the result has been that the indoor livestock units have congregated on the western side of the country, leaving the drier, flatter, sunnier eastern side to the 'barley barons.'

The intensive indoor rearing of livestock is what is sometimes called 'factory farming.' The first creatures to find themselves 'industrialized' in this way were hens. Egg and chicken production had been a sideline for many farmers. The traditional picture of a farmyard shows a few birds scratching around the yard. They lived on scraps of food from the house, on spilled grain and on such grass and green stuff as they could find. They provided the farmer's household with eggs and with birds for the table, with some over to sell, but economically they were of no great importance. Specialist poultry farmers kept their birds out of doors and folded them across their fields, surrounding them with a moveable fence that confined them to a restricted area that also contained their sleeping quarters. Every so often the houses, fences and all would be moved to expose an area of fresh grass and to allow the area they left to recover, fertilized with their droppings. When they moved indoors they entered one of two main systems. Most were – and are – kept in cages in tiered rows. These are battery hens. Their food reaches them along a trough at the front of the cage and the eggs they lay roll down, out of the cage, for easy collection. Table birds are raised in deep litter, or broiler houses. They are free to move about – or they would be were they less crowded. They are fed from hoppers.

The intensification of the poultry industry began in earnest in about 1950 and since 1954 it has grown rapidly. In that year there were some 20 million broilers produced, in 1957, 56 million, in 1959, 108 million, in 1960, 142 million, and in 1969 we produced 230 million birds. Chicken moved from a high-priced luxury food item into the cheapest form of meat.

The next animals to move indoors were calves being reared for veal. The British Quality Veal Producers' Association has stated that in 1969 there were around 40,000 animals being reared for veal in Britain. Then pigs moved inside. Some 11.5 million pigs were killed in Britain in 1969. To a lesser extent dairy cattle have been brought indoors and attempts, so far with no great success, have been made to bring in sheep.

In 1964, a book called *Animal Machines,* by Ruth Harrison, described intensive livestock rearing. She coined the term 'factory farming.' The book attracted a great deal of publicity and it aroused a passionate response from consumers who had had no idea of the methods that were being employed in the production of their food. Mrs. Harrison's criticisms of these methods were concerned mostly with the cruelty involved – and there is no doubt whatever that they are hideously cruel. The government appointed a committee to study the question, under the chairmanship of Professor Rogers Brambell. The report of this committee, the Brambell Report, contained specific recommendations relating to the conditions under which farm livestock should be confined. The Report appeared at the end of 1965 and so far its recommendations have not been implemented. The cruelties continue.

Those who campaign against 'factory farming' are usually challenged with the argument that the world food problem is so acute that we must not be too scrupulous about the methods we use to produce food. People must be fed. 'Factory farming' may be unpleasant but it is highly productive and without it millions more would starve. It is appropriate, therefore, to consider intensive, indoor rearing of livestock in relation to world food supplies. Will the broiler houses feed the world? If people can be fed, they must be fed.

First, let us make it quite clear that at the present time 'factory farms' make no positive contribution whatever to world food supplies. Britain exports no broiler chickens to India and although there is trade in animal products it takes place among the developed countries themselves. The third world is not a recipient.

Could it be? Could British farmers export animal products to developing countries? Probably they could. One of the effects of the high levels of productivity achieved by livestock farmers is that they have glutted their markets. A year or two ago it was said that if you wanted to become rich in Britain the thing to do was to buy a very small piece of land, build a cold storage warehouse, rent space in it to the poultry farmers, close and lock the door as soon as it was filled, and then sit quietly at home drawing rent. You would not be required to open the door ever again! Certainly it is true that turkeys and capons that appear in the supermarket at Christmas go back into store and come out again at Easter. There is a limit to the amount of poultry people can eat. In spite of expensive advertising campaigns, the British are still not eating enough eggs. So there is food to spare.

Unfortunately, it is very expensive food, produced in a high-wage, high-cost industrial system. It is always more expensive to buy a finished product than to buy the raw materials and produce it yourself. All meat products are 'finished' in this sense and since the developing countries find it difficult to afford to buy grains – the raw materials – there is no chance whatever that they could buy animal products. So, although British farmers would be glad to export, no one can afford to buy from them, except in the other developed countries, all of which have their own intensive systems – some more intensive than those in Britain – and are in the same boat.

Could the developing countries not introduce intensive systems themselves and so increase the proportion of animal protein in their diets? Again, although 'factory farming' is a misleading term in some ways, it is less so when applied to a developing country. It is based on buildings, environmental controls, automation. The large-scale 'factory farm' is a sophisticated product of an industrial society and the capital required to establish it is high, even by British standards. Thus the first question that needs an answer is, if intensive units are to be established in developing countries, who is to pay for them? Could government funds be invested in this direction? Perhaps they could, but even so it would not be cheap. The economics of intensive livestock farming are misleading. In Britain, over a ten year period, the government paid over £800 million in subsidies to farmers that were of direct benefit to the intensive livestock farmers,* so placing the enterprise in the position of a

* According to Prof. J. R. Bellerby in *Factory Farming*, published by the British Association for the Advancement of Science, 1970.

developing industry receiving active official encouragement. Even then, the intensive unit will function only if it receives the inputs it needs. These include power and water and they also include food. This can be produced only by arable farmers – or fishermen, but we will come to them later – and so the arable farms must be made healthy and productive first. We have gone full circle. The first need is to improve productivity on arable farms.

Once the inputs are available so that livestock production can increase, is this the way to do it? Speaking at a session of the 1968 annual conference of the British Association for the Advancement of Science that was devoted to 'factory farming,' Dr. N. A. Mujumdar, an adviser to the Bank of Zambia, said:

'An overall judgement of the situation suggests that the conditions in Zambia do *not* warrant the promotion of indoor rearing of livestock.' He went on to give some of the reasons: '... because of the very low density of population, the land labour ratio does not warrant such an approach to livestock development. Nor is there any need to economize labour; in fact, urban unemployment is increasing ... The most important problem in this context is therefore to train Zambian farmers. Even from the purely farm technology point of view, training Zambian farmers in the traditional livestock rearing methods could perhaps be much easier than training them in the techniques of indoor livestock rearing.' A Ghanaian speaker at the same conference stated that the situation described as applying to Zambia also applied to Ghana. F. D. T. Good, a British veterinary surgeon with wide experience in several parts of Africa, agreed. 'No advantage can come from keeping cattle and other large stock indoors in view of the abundance of labour and land,' he said, although he considered there might be advantages, in some places and under some circumstances, in the intensive production of poultry and smaller animals. In its summary of the IWP, the FAO calls for an increase in meat production in developing countries, but it makes no mention of intensive systems as being relevant. However, it does underline the input requirements that must be taken into account in any attempt to increase production of animals. If the developing countries, it says, could increase their poultry meat production by 5 per cent, the gap between demand and supply would be closed. However:

'To achieve such a high poultry and pig production, however, would require fulfilment of several important conditions. The livestock proposals in the regional studies would require over 80 million tons of concentrate feed by 1985 compared with 30 million tons in 1962 . . . Some 70 to 80 per cent of this would need to consist of cereals and their by-products, and therefore feed-grain requirements in developing countries by 1985 could reach 80 to 90 million tons compared with just over 20 million tons in 1962.'

There seems little doubt that 'factory farms' export no food to developing countries, nor will they do so, and that 'factory farming' is wholly irrelevant to the needs of the third world. If it is considered desirable to increase meat production, and if the established farming systems provide the inputs to make this possible, then there are alternative methods more suited to the immediate needs of the people.

If it contributes nothing positive to the world food problem, does 'factory farming' bear any relationship to it at all? Unfortunately, it does. It has a very definite adverse effect.

Intensive livestock rearing is a phenomenon of the developed countries and as a system it has added to the availability of meat products within those countries. It has achieved levels of production that would be impossible within the food production capability of those countries themselves. There is not one of them that has land enough to grow cereals for feedingstuffs on the scale that is demanded by their livestock farmers. The balance is made up by imports, and the only possible source of such imports is the third world. To a large extent, then, intensive livestock farming in Western Europe and the US is subsidized by the developing countries. The imports arrive in the form of fishery products and oilseed cakes and meal. Prof. Georg Borgstrom* has described the position with regard to fishery products:

'With the aid of West European and United States capital, Peru has grossly expanded its fisheries into the rich Peruvian current (traditionally called the Humboldt Current) off the Pacific coast of South America. It now almost surpasses in volume Europe's entire

* *Too Many*, Macmillan, 1970, p. 235.

sea catch (9.8 million metric tons in 1967*). Chile is following suit, rapidly expanding its fish-meal industry, but on a more modest scale. Most of the invested capital is West European or American; some is national. Close to half of the protein taken from the Pacific (1966–8) emanates from these new fisheries in Peru and Chile – quite a remarkable accomplishment. Practically the entire catch is converted to animal feed in fish-meal factories. A major part (80 per cent) of the fish riches of the African Benguela Current is in a similar way channelled into reduction.

'In both cases almost the entire output of meal and oil goes to Western Europe (approximately two-thirds) and the United States (slightly less than one-third) with a minor part to Japan, where it constitutes a significant base for the budding animal production. The oil serves the European margarine industry. The impressive increase of animal production in postwar Europe is to a substantial degree due to this gigantic influx of first-rate animal protein, which enters into agricultural production almost through the back door.'

So large is the volume of fish-meal imported from Peru and Chile, that were the whole of this protein to be diverted to the home market, the protein content of the diet of all the people throughout the whole of South America would be raised to the level of that of Southern Europe. In 1966–68, says Borgstrom, the amount of protein exported from the region exceeded by 50 per cent the total meat protein production of Latin America, and it was three times greater than the milk production.

Yet Latin America is not the only source. The fish-meal exported from Africa to the developed countries could, if consumed by Africans, halve the protein shortage throughout the continent.

When vegetable oils are extracted from ground nuts, sunflower seeds and soya beans, the residue is a pressed cake, rich in protein. The world's oilseed cake protein output exceeds that of the world fisheries. Most of it goes to the developed countries for cattle feed. There is nothing new in this. We have been importing cattle feed for many years. At the present time, however, of the world's total trade in oilseed cakes and meal, 93 per cent goes to Western Europe. Of course, oilseed cake is not edible to humans, although a large

* We will consider more recent figures for fish catches, and the distribution of the proceeds, in the next chapter.

amount of research is devoted to making it so. If it were, the present volume of production is such that the protein available throughout the developing countries might double.

So we see that far from being unavoidable if the starving millions of the world are to be fed, 'factory farming' takes valuable food from hungry people and gives it to well-fed people – and on a colossal scale. It does not even feed the poor in many of the developed countries, for although they are cheaper than they would be were all animals reared out of doors, extensively, animal products are still expensive and large numbers among the poorest sections of the population cannot afford to buy them. Broiler chicken may have become the cheapest form of meat on the market, but it remains beyond the purse of many an old-age pensioner.

Clearly, this situation cannot continue. A point is bound to be reached at which the hungry realize that food is being taken from them to indulge the rich. The price of imported animal feed will rise sharply, to a realistic level, or supplies may cease altogether. When this happens, the intensive livestock farmers will be able to produce no more than can be fed from their own arable acres and it will be interesting to compare productivity with that which was achieved under traditional systems of husbandry.

The intensive livestock industry may collapse. In point of fact, its collapse may come even sooner, because of contradictions inherent within the system itself. We have seen that there is a danger of saturating the market, leading to gluts, dumping and bankruptcies. Producers are compelled to lower their prices if they are to remain competitive. As their profit margins approach their costs, setbacks may prove disastrous. The immediate solution is usually to seek economies of scale. It is the largest producers that survive each time. So the industry tends toward monopolies. The situation may resolve itself in time, but if it continues on its present path it will do so through the exclusion altogether of the small producer. All our meat products would emanate from the units of a very small number of very rich and powerful agro-industrialists. Even then, profit margins per unit production will remain small and the production process will require to be as cheap as possible. This will result in a certain economic instability related to biological instability.

The biological problems are more troublesome than the purely economic ones, although an economic solution might be found for the first of these.

The deterioration of soil structure is due to a reduction in the amount of organic matter it contains, and the reason for this is that as animals have been removed from the land, the land has been deprived of their urine and faeces. These are still produced, of course, and in larger quantities than ever, for the number of animals has increased. Now, however, the wastes are produced inside indoor units and instead of being regarded as a raw material – a source of fertility – they are an embarrassment. The new stock farmers have little or no land to which to return them. Indeed, almost by definition, they have less land than could accommodate their flocks or herds under an extensive regime.

At first they did what we have nearly always done with our sewage – they poured it into the rivers. There, being rich in plant nutrients, particularly nitrogen, it caused eutrophication. Legislation was enacted to control the pollution of rivers and slowly, in the face of vigorous opposition, sometimes from the courts themselves, the river authorities began to enforce it. Under existing laws, farm wastes must be treated to a satisfactory level before they are discharged. Penalties are pitifully small and there are many cases of small industrialists finding it cheaper to continue to discharge their effluent and pay the fines, rather than instal treatment plant. (In November 1971, Mr. Peter Walker, Minister for the Environment, announced stiffer penalties for pollution.) Farmers, however, began to seek alternatives. If they could not discharge into the rivers directly, the most obvious thing to do was to discharge into the public sewerage system, and so pass on the problem to the local authority. The local authority would not stand for this. A medium-sized livestock unit produces as much sewage as a small town and sometimes, especially in the case of pigs, the effluent is highly potent. The sewage works were not able to cope with this increase. If the local authority was to accept the effluent, new plant would have to be installed and the cost would have to be passed on to the ratepayers. Since the units were generally in rural or semi-rural areas, the farmers and their workers were also the ratepayers, so that although the treatment cost might be subsidized by their own employees, the situation was not altogether satisfactory. Yet the cost of installing plant on the farm to treat their own effluent to the standard required by the river authorities was often prohibitive and always there was the risk that if other industries, or other intensive livestock units, moved into the area, the permissible contribution of pollutants from each would be correspondingly smaller, calling for treatment to higher standards. In 1968 it was estimated by the National Farmers' Union that a pig farmer with a moderately sized unit might have to find £15,000 not to mention operating costs. In spite of Britain's elaborate agricultural support system, no major grants were allowable to help toward this expense.

This left two alternatives, and both are in common use today. The first was to build large lagoons into which the effluent could be poured and kept. As each lagoon filled, another was opened. Sometimes it might be possible to quick-dry the effluent and make it into an organically-based fertilizer. This is done to some extent with poultry manure. The other alternative was to pour the effluent on to whatever land was available. As it filtered through the soil the solids would be trapped and by the time it reached the river or groundwater the liquid would be at least partially purified. This process kills the land more effectively even than would depriving it of its organic matter, for the rate of bacterial decomposition is so high that the natural organic content of the soil is oxidized away.

The problem of effluent disposal could be solved if a way could be found to transport it economically from one side of the country to the other. This is more difficult than it may seem, because it is extremely bulky. This means costs are high, so that the prospective recipient farmer cannot afford to pay for it. To him it must compete with artificial fertilizers, for he tends to think in terms of nutrient rather than organic matter content. While fertilizers are subsidized, organic manures are not. A revision of the support system might help to a solution.

Regarding animal wastes as anything but a valuable raw material is little short of insane and present treatment – or non-treatment – causes serious pollution and considerable offence. This is becoming well-known in rural areas and when a new unit is planned, local opposition is often strenuous. Yet there is another problem which poses a direct and grave threat to human health.

If animals are housed in large numbers in close proximity to one another, any disease that strikes one may quickly develop into an epidemic. Since the farmer may have much of his capital invested in his stock, which represents his sole source of income, the implications of epidemic disease are serious indeed. Antibiotics are available to treat most diseases of livestock, but the danger remains that should disease strike it could spread through the unit too rapidly for any therapy to be effective. It makes sense, therefore, to dose the animals with drugs to prevent disease, rather than to attempt, probably too late, to cure it. So the prophylactic use of antibiotics, in small doses, became routine. It was discovered, too, that minute doses added to feed acted as a growth stimulant. So it became almost universal practice to include antibiotics in animal feeds.

Antibiotics are rather like insecticides. They aim to kill the target species – in this case a micro-organism – but unless the size of the dose is large enough not all of the disease organisms will be killed. As with insects, the survivors will include individuals that are naturally resistant, and so with repeated doses a resistant population will evolve. Since micro-organisms reproduce much more rapidly than do insects, resistant populations will appear more quickly. If the drugs are used only for therapeutic purposes, this is not likely to create a serious problem, because the only individuals to be treated are those actually suffering from illness and the treatment is not repeated after recovery. The drugs are used very specifically and with great care and discrimination.

When they were introduced into animal husbandry a new situation developed. Now small doses were being introduced repeatedly into the environment and the appearance of resistant strains became inevitable. The British Veterinary Association, in its evidence to the Brambell Committee in 1964, warned of the danger to human health from this source. Between 1957 and 1960, the Ministry of Agriculture's laboratories at Weybridge found that in samples of *E. coli* bacteria, which is a harmless species common in man, taken from the gut of farm animals, the percentage that was resistant to most of the antibiotics in common use rose from 3.5 to 63.2. In 1963, *Salmonella typhimurium* type 29 began to acquire resistance and by 1965 is was fully resistant to most antibiotics. Between 1965 and 1967, three persons died and ten were made dangerously ill by resistant strains of *S. typhimurium*. Since then there have been other cases of food poisoning that did not respond to routine treatment.

Once a microbial population acquires resistance to a particular drug it does not lose it, and the resistance is transmitted genetically from one generation to the next. In this way strains can appear that are resistant to all the antibiotics they have encountered – and this includes most of those in common use.

This is called multiple resistance, but in 1967 evidence began to appear of multiple-multiple resistance. It appeared that resistant strains of organisms could 'infect' non-resistant strains, so that resistance was transmitted laterally, as it were, as well as vertically. Then it was learned that not only was resistance transmissible in this way between strains of the same species, but it could also pass from one species to another. Once a strain of organisms becomes resistant to antibiotics, all the other organisms, of whatever species, with which it comes into contact, may also acquire resistance. The threat to human health is obvious, but the most serious aspect relates to the risk of typhus. There is only one drug that is really effective against typhus – chloramphenicol. Chloramphenicol is one of the antibiotics used in animal husbandry. If the typhus bacillus should become resistant, then any outbreak of the disease caused by it could not be treated.

Nor does the problem end there. Whenever antibiotics are used, they kill more than the target species. The large majority of the microflora and fauna in the bodies of animals serves a valuable, sometimes essential function. Part of this function involves opposing invasion by pathogenic organisms. So, when antibiotics are used, the quantities administered must be such as to take account of the fact that allies are being killed as well as the enemy. The drug must take over full responsibility. If the enemies are resistant but the allies are not – and this is quite possible, since the allies have developed within the body, protected from exposure to antibiotics and resistant micro-organisms alike – then the effect of the therapy will be to make the illness worse, while at the same time interrupting the other bodily processes for which the gut flora and fauna were responsible. The therapeutic drugs that have achieved so much in the treatment of infectious disease may be turned into weapons against us. At best we are debasing the currency.

In July 1968, a committee was appointed to examine the use of drugs in livestock husbandry. Its report, The Swann Report, appeared in 1969 and it recognized the existence of resistance, multiple resistance and multiple-multiple resistance. It recommended in principal that drugs should be used only therapeutically, unless they were drugs that were not used to treat illness in animals or humans. It listed the antibiotics whose use should be restricted. It did not advise an outright ban, for it recognized the serious economic implications within the livestock industry.

The opposition to the Report was immediate and vociferous. The sale of antibiotics to farmers was of considerable importance to the pharmaceutical industry. In January 1970, a symposium on 'The implications of infective drug resistance' was held at the Royal Society of Medicine, in London. The invitations said the symposium was to be held at and by courtesy of the Royal Society of Medicine. Anthony Tucker, writing in *The Guardian* (19 January 1970) smelled a rat:

'A member of the administrative staff of the Royal Society of Medicine, before she realized that she might be quoted, said that the Society had "hoped that its section of Comparative Medicine would take part." The haste with which the meeting was convened made this impossible.'

'Yet Cyanamid has brought two of the speakers from the US together with their director of research and other staff research scientists who are to participate in discussions from the floor. This is clearly not an operation which could have been planned and brought to fruition in a few days. Strangely, Cyanamid's director of research arrived in Britain a week ahead of time, some days before very late invitations were sent out to British academic experts in this highly specialized field.'

It transpired that the symposium had been sponsored and arranged by one of the leading pharmaceutical companies and the speakers were, not surprisingly, strongly opposed to the Swann recommendations.

Like the recommendations of the Brambell Committee, the Swann recommendations have never been implemented fully, although there have been some restrictions on the use of certain drugs without prescription.

If antibiotics were withdrawn, could intensive livestock farming continue? There is considerable doubt.

It appears, then, that 'factory farming' has an overall disadvantageous effect on food availability in the developing countries. It is highly improbable that it will be introduced in those countries because of the high investment required and because it is inappropriate to the needs of the people in the regions concerned. So far as the world food problem is concerned, we would be better off without it.

Its future in the developed countries is uncertain. The biological problems it faces in disposing of its wastes and in controlling its ever-present disease risk without the use of drugs may be insuperable. It may founder because ecologically it is unsound.

If it should survive its immediate difficulties, it will face others as soon as the demand for food in developing countries creates sufficient pressure to necessitate the diversion of protein from export to the home market. The competition for available fish-meal and oilseed cake will bring sharp price increases as the supply dwindles, and eventually the supply may cease altogether. At that point the livestock industries of the developed countries will find themselves contained within the resources of their own territories and there is no reason to suppose that in such circumstances they can provide a satisfactory alternative to more traditional methods of husbandry. At the same time, the condition of the soil in the arable growing areas is such that it may be found necessary to revert to grassland farming and this will encourage a return to the raising of livestock out of doors.

All in all, it is very unlikely that the present intensive livestock systems can survive for more than a few years more.

10

Wishful thinking

We are caught in a race between population and resources. We are concerned here with exploring the final limits of this planet's capacity to support human life. How many of us must there be before there is a breakdown in the life-support systems on which we depend?

Probably this figure is unknowable, partly at least because before it can be calculated we must agree among ourselves about what we mean by 'life.' It has been said, for example, that those who concern themselves above all with nutritional quality in agricultural products should remember that discussions about increasing food production are conducted in a context of the hungry – people who will die next week if they cannot be fed. So far as they are concerned, a pre-occupation with nutritional quality must seem irrelevant. Almost any food will do. Yet is this all that we hope to achieve? Will we be satisfied if people can be kept alive, somehow, for a further week? Is this what it is all about? If it is not, if we aim for more than a hand-to-mouth lurch from crisis to crisis, then we must take account of the quality of life and the nutritional quality of food, which is related to human health, assumes immediate importance.

Nevertheless, even before we can calculate how many people the world can support at the barest subsistence level, we must examine one or two ways more in which food is, or might be, made available, and we must consider what might be achieved by redistribution of existing populations.

The anti-populationists point out that there remain areas of the world that are under-populated. This is true, although with populations rising it is little comfort to be told that until the last corner is filled there is no problem! Indeed, the term 'population control' is often misunderstood to mean no more than 'population limitation.' In fact it means relating population to the carrying capacity of the environment. There are areas that would benefit from a rather larger population. Some of these areas are located in developing countries. Ethiopia, for example, regards itself as under-populated with 59 persons per sq. mile. At the plenary session on population at the Second World Food Congress, an Ethiopian delegate said his country needs more people. What he went on to say, however, was that the country's own population growth must supply these people. He was rejecting a suggestion that people from over-populated areas might be resettled in areas that were less crowded. The suggestion, incidentally, had been made by an Australian!

The Ethiopian's point was a good and a valid one. If people are moved in large numbers from one part of the world to another, problems are created that may be even more intractible than those the move was meant to solve. The further removed the people are from western culture the more difficult these problems are likely to become.

Western Europe and America have imposed their culture and its values on a large part of the world, so that in theory at least, a citizen of any European country, or a North American, may feel at home in a number of places. He will find the legal system broadly similar, based on similar concepts of justice, he will find schools, universities and other academic institutions teaching different subjects, perhaps, but basing their teaching on similar ideas and with objectives similar to those of the schools he left behind him. He will be able to recognize a school as a school. Industrial production, marketing and commerce generally will operate in a somewhat similar fashion to those he is used to. Yet even then, population movements are never easy. Immigrants tend to be feared and resented, small differences in custom assume large proportions and adjustment is difficult. Imagine the plight of a group of middle-aged English workmen and their families moved, shall we say, from the West Midlands to Greece, where they must learn to speak and read Greek, to work and live as though they were Greeks. Would the Greek host community accept them? Would they themselves be able to make the adjustment? If they did, would they retain any of the characteristics that had made them English? Should they? After several centuries, the European population of the United States has not learned to live in peace and as equals with the non-European section of the population. The British pride themselves on their tolerance, yet their attitude to immigrants from Asia, Africa and the West Indies suggests they are as racialist as any other nation. Yet all these groups accept, to a greater or lesser degree, similar values and cultural practices. They have been 'westernized.'

In more primitive communities people tend to be more conservative and more closely tied to a particular locality and particular customs. When the Volta Dam was built in Ghana, large numbers of villagers had to be moved to new settlements. All kinds of difficulties arose. Some people could be buried only in a particular place, some tribes would not live next to certain other tribes, and many, who were polygamous, could not live in the concrete houses built for them, because the new houses provided accommodation for only one wife.

With the best of intentions, a tribe of Amerindians living on the borders of Colombia and Venezuela were provided with new houses to replace the communal 'bohios' in which they were used to living, the whole tribe under one roof. The culture of the tribe collapsed; it had needed the communal dwelling.

Canadian Eskimos, resettled to be nearer an available food supply after over-hunting had caused the caribou herds to change their migration routes, so removing the Eskimos' main food source, left and chose starvation rather than share a small island in a lake with a tribe of Indians that they feared.

Moving whole populations – even small ones – is fraught with danger and difficulty, and it is very expensive. The rejection by the Ethiopian of any suggestion that immigrants from other countries should be settled in his country was less unreasonable than at first it might seem.

Even supposing that the cultural difficulties can be overcome – and if it is essential to do so, no doubt they can be – how many under-populated areas are there left? The first that springs to mind is Australia, but the Australians themselves are beginning to realize that they are approaching their own limits. Australia cannot accommodate many more new arrivals. Canada has an overall population density of five people per sq. mile, but most of these people live within 500 miles of the US border. Most of Canada is in or very close to the Arctic. Large parts of it are uninhabitable. The areas that are pleasant to live in are more densely populated. New Brunswick has twenty-two people per sq. mile, Nova Scotia thirty-eight, and Prince Edward Island fifty-five. There is room for more people in Canada, but not all that many more, and the Canadian Government, which is probably the most enlightened government in the world with regard to environmental issues, demands that immigrants be people who can contribute to the quality of life and prosperity of existing Canadians. That reduces the field.

No, we cannot escape by settling new territories. It worked in the seventeenth, eighteenth and nineteenth centuries, but now there are few new territories left and those that are will soon be filled. In any case, as we have seen already, almost the whole of the planet's cultivable area is growing food already, so resettling people is unlikely to add much to world food production, even though it might take people from areas of scarcity to areas of plenty, so solving distribution problems. Equally, they might remain at home and have the food brought to them. This would be simpler and cheaper, although we have had to reject the importing of food as a solution to the shortages in the developing countries. Even so, would there be enough?

The technophiles have suggested that soon we will be able to export people to other planets. This is the most fanciful suggestion of all and it would not be worth even a note were it not for the fact that so many people in the developed countries today have such faith in technology and its ability to find solutions to any of the problems it has created that they believe the fairytale. They dream of giant spaceships conveying millions of surplus people (who decides who is surplus?) far away, through deep space . . . to where? The other planets in the solar system are uninhabitable unless complete life-support systems are installed on them. We lack the technology to travel beyond the solar system. Actually, at the present time, we lack the technological capacity to convey passengers any further than the moon, but let it pass.

All kinds of possibilities have been suggested. One is that living organisms be injected into the atmosphere of Venus from an unmanned craft. This, it is said, would speed up the evolution of that planet along the path the earth has taken. The idea is attractive.

Unicellular algae would break down carbon dioxide in the Venerial atmosphere and would liberate oxygen. The algae would multiply, the process would accelerate. As the carbon dioxide content of the atmosphere was reduced it would be possible for reflected radiation to leave the planet, which would grow cooler. Eventually it would be cool enough for water to condense. A cool surface, an atmosphere with oxygen, and plants and water, and it is quite possible that evolution might get under way with a bang. The planet might become a very pleasant place inded for the first human colonists, a veritable Garden of Eden. The entire process might be telescoped from hundreds of thousands of millions of years to one or two million. Until then, how shall we pass our time?

Even were the technology up to it, what would it cost? Professor Garrett Hardin of the University of California, has calculated that if the people of the United States were to reduce their standard of living to 18 per cent of its present level, and if they set aside all the money that they saved for this one project, then in a year they could accumulate enough to transport into space the world's increase of population *for one day*. Even if we could export all of the world's additional population in some miraculous way, Professor Paul R. Ehrlich, of Stanford University, has worked out that it would take about 200 years to fill all the planets in the solar system to overflowing. We cannot escape the problem. We must solve it here. But how?

Can food production be increased in any way other than by conventional agriculture? What about fish? The oceans cover most of the surface of the planet and it is easy to imagine that they are bottomless and their bounty inexhaustible. And, of course, fish is already a major source of protein, either for direct human consumption or, as we saw in our consideration of 'factory farming,' for reprocessing through farm livestock.

The FAO grows quite lyrical in its optimism. In one of its reports issued to mark the end of its first quarter century, it says:

'By the year 2000 man may have explored all the ocean depths, mined the seabed for its mineral wealth and farmed the coastal seas for plant and animal food.

'Such activities may seem less spectacular than the expected conquest of the planets. But they may well prove more vital, especially as regards food. For as mankind races towards the 21st century, the problem of wresting proper nourishment from a shrinking planet becomes increasingly urgent.'

'The sea, covering 70 per cent of the earth's surface, is an immense treasure-trove of protein-rich food, producing each year thousands of millions of tons of animals and plants from whales to plankton. Moreover the resource, a miracle of ecological balance, is entirely self-renewing, requiring no intervention by man. As custodian and chief beneficiary of the resource, he has only to harvest it . . .'

The international fishing industry was reduced during the Second World War, but after the war production increased rapidly. In 1950 the 1938 total catch of marine and freshwater fish – 21 million metric tons – was passed. In 1960, 40 million tons of fish were caught, in 1964, 53 million and in 1968, 64 million metric tons.

The increases were achieved by heavy capital investments and considerable technological improvements. The old fishing boats were a thing of the past as the millions of holidaymakers who visit the dying fishing villages of England and Scotland must have noticed. They were replaced by big, deep-sea trawlers with refrigerated holds, working in fleets to supply factory ships of 20,000 tons or more. They introduced radar and sonar and computers which can calculate the strategy the skipper should follow to pursue his shoal. They fished further and in deeper water. The large nets now may be more than a quarter of a mile long and more than 500ft deep. A large, modern seine net could hold the Colosseum in Rome. There are trawl nets half the size of a football field.

In spite of these advances, the food supply of the developing countries has not increased greatly. The reason has been that they have been too poor to invest in the sophisticated equipment required. The big fishing nations are the developed countries – Japan, the USSR, Norway, Denmark, Spain, South Africa and the USA. Mainland China has also increased its fisheries to a considerable degree, probably because it possesses the centralized control necessary to divert resources into what it regards as a valuable area. So has Peru, but the Peruvians do not eat the fish they catch. At the present time Peru is working through the FAO to increase its catch, so that it may have some for domestic consumption.

Several of the major fishing nations operate bases outside their homelands, but even then the fish intended for direct consumption returns home and does not reach the indigenous market. Fish that enters the world market is canned or reduced to animal feedingstuffs or pet foods for use in the developed countries.

The world fishing industry will continue to expand for some time yet. It has to, because of the large capital investments that have been made already in ships, equipment and processing plant and which must be recouped. At the same time, investment in ships and equipment must continue to increase because at present it does not match the capacity of the processing plants. The plants cannot be fully profitable unless they work at capacity, so there must be more ships. The proportion of the world fish catch that is processed into animal feeds now exceeds one-third. The hungry nations are bypassed in this traffic. While, in 1967, some 40 per cent of the world's catch was made into fish-meal, only 17 per cent went to feed the developing countries.

There is a precedent for this, of course. The expansion of population in Europe was made possible by opening up and exploiting sparsely populated food-producing areas in North America, Australasia and, to a lesser extent, in Africa and Asia. In order to support a further expansion, we have moved into the oceans. Fisheries may be feeding the world, but when we say this we are using the word 'world' in the way that the Romans used it. The fish are feeding our world. The poor and hungry areas are on the outside.

Can it continue? In 1969, for the first time in twenty-five years, the world fish catch dropped by 2 per cent, to 63 million metric tons. The proportion used in fish-meal was also slightly down, from the 1968 figure, to 35.8 per cent, but even so it exceeded the levels for all previous years. The top ten fishing nations in 1969, in descending order, were Japan, Canada, Norway, Peru, Denmark and the Faroes, USA, Iceland, USSR, Spain and South Africa.

Why was the yield down? The FAO has expressed fears about over-fishing, but at the same time it estimates, paradoxically, that improved technology, permitting fishing at deeper levels, could bring a maximum sustainable catch of around 100 million tons a year. At the same time, the variety of fish eaten could be widened. Indeed, ideas of which fish are edible is so much a matter of cultural preference that many fish are rejected, or processed, that are considered highly edible elsewhere. The main ingredient of fish-meal production in Norway and Iceland, for example, is herring, and the anchoveta that goes to make Peruvian fish-meal was dried and distributed throughout the whole empire in Inca days.

The fishing nations are too far committed to allow themselves to be deterred by a drop in the catch for one year. The USSR alone plans to increase its share from 9 to 14 million tons a year by 1975.

Is it possible? The increasing pollution of the oceans is reducing the planktonic food supply and this may affect the ability of fish populations to recover after they have been harvested. As fishing moves into deeper and more remote waters, is it really seeking larger shoals? What has happened to those inshore? Why did the 'cod war' between Iceland and Britain become so heated? Were fishermen from Britain really taking so much from the inshore Icelandic fishermen? Or are there less fish in the sea than there were? Are we 'mining' the oceans? If we are, then the FAO estimate of sustainable yield may be far too high. If pollution is having a cumulative adverse effect, then fish populations may be declining from two separate causes and the situation may be worse still. There are scientists who estimate that long before the end of the century all the oceans will be dead and some of the shallower seas that are almost completely land-locked, such as the Baltic, the Mediterranean, the North Sea and the Irish Sea are already in serious trouble.

Obviously there must be a limit to the hunting of a wild population beyond which the population will be destroyed. This is the reason for restrictions on all forms of hunting for sport in most countries. If we are too greedy we will spoil things for everyone, including ourselves. While we can behave with some sense of responsibility within our own frontiers, however, our behaviour internationally gives little ground for optimism. Certain species of whale are on the point of becoming extinct because of over-hunting. Marine biologists have warned for many years that the populations were being over-exploited, but their warnings were ignored. They achieved representation at the highest levels, they won the ears of the international committees that set whaling limits, and still they were ignored. The reason, probably, is that the whales and, indeed, the oceans, belong to no one. They are a common resource and it is the fate of common resources that they will be destroyed.

Professor Garrett Hardin once described this process very graphically in an essay, that has since become famous, 'The Tragedy of the Commons.' A common is a natural resource that belongs to a community – village or global, the principle is the same – rather than to any individual or identifiable group of individuals. We can illustrate what happens by considering a piece of common land in which the inhabitants of a village all share the right to graze their cattle. Each man in the village has one cow, and there is enough grazing for all. However, since the food is free, the man who manages to buy a second cow must invest capital but he will have little in the way of operating cost and his income will double. The advantage to him in acquiring a second animal is obvious. As soon as he has one all the other villagers see that he is now richer than they are. In an effort to match him, and to prevent him from using his economic power to assert political domination over them, they, too, must each acquire a second cow. The situation re-stabilizes. Everyone is better off, but there are now twice as many cows as there were before. Another man now decides to buy a third cow. The advantage to him is the same as before. His income increases by 50 per cent and, again, he has a low operating cost. Moreover, because of the additional income provided by the second cow, he has less difficulty in finding capital to invest. However, the addition of this one extra cow overloads the land. It is being over-grazed. There is not enough food for the extra cow. However, this does not affect its owner. What happens is that all of the cows receive slightly less food. In other words, while the advantage of adding another animal accrues entirely to its owner, the disadvantage is shared equally among the community as a whole. Someone else acquires a third cow, and someone else again. The quality of the pasture declines, but still the advantage to the individual outweighs the disadvantage, until the system breaks under the strain. The land fails to regenerate, the grass dies and all of the cattle are left without food. At every stage up to the last, the advantage to the individual was immediate and obvious, all the disadvantages being shared. In the latter stages, moreover, as income from each animal begins to decline, the response may be to step up investment, to add more animals to compensate for the lower yields from each. The process is inexorable and history provides very many examples. If the analogy is valid, have we any real grounds for supposing that the seas will not be fished to destruction?

Meanwhile, is there any reason to hope that increasing the fish catch will benefit the hungry? India has a large fishing fleet which it hopes to enlarge further, provided financial and technical aid is forthcoming from its competitors! The one exception is inland fisheries, which are contained within national boundaries although, even then, there may be competition in rivers that cross frontiers. On the whole, it is likely that the level of investment required to secure the yields that the FAO anticipates will effectively exclude the developing countries from harvesting the deep oceans, while river and estuarine pollution and ecological disruption caused by irrigation development inland, to which must be added the increasing activity of the inshore fishermen, will deplete inshore fisheries. The prospect is not bright.

The cultivation of fish may offer some possibilities. This is 'fish farming' and it has created a certain amount of excitement in the west, as though it were new. In fact the British Museum contains a text book on fish farming that was written in China in 850 BC! For centuries the paddy fields of the Far East have been used to raise fish before they were drained, so bringing two harvests from one piece of land.

There is a certain amount of research into the possibilities of fish cultivation being conducted in the developed countries. In Russia they have developed a hybrid, called the 'bester,' that may be husbanded to take over from the migratory white sturgeon as the main source of caviare. The sturgeon has been reduced in numbers by ecological changes that have been wrought by major hydro-electric schemes and by pollution in the rivers where it spawns. There are many technical problems, however, and progress is slow. Experiments are also being conducted in the USSR into breeding fish that will live in over-enriched (eutrophicated) water. The fish can be harvested and they will reduce weed growth. Again, there are serious technical difficulties, some of which we have met before. One, for example, is the risk of epidemic disease among densely stocked shoals, the problem that was encountered by 'factory farmers.' The Biochemical Fisheries Research Institute of Aberdeen University has had some success at the immunization of fish colonies, but the cost of immunizing each of the many thousands of individuals that would be stocked in a commercial unit may prove prohibitive.

It is possible that in years to come the controlled cultivation of fish may bring dividends. We should remember, however, that fresh fish require fresh water and that they will have to compete for this with other users and that, whether they are to be cultivated in fresh or sea water, they will require feeding. The most likely basic nutrient might be artificial fertilizer, which would stimulate the growth of the microflora that are at the base of the aquatic biotic pyramid.

There are other ways of producing protein, which may or may not involve the use of domestic animals as 'converters.' Even when they are managed intensively, animals are not very efficient as manufacturers of protein edible for humans, especially if they are consuming a food that could be eaten by humans directly, or a food grown on land that could be devoted to a high-protein vegetable crop for human consumption. Professor W. Holmes, of the University of London, in a paper presented to the symposium on Potential Crop Production in Britain, contrasted yields from different farming enterprises in terms of their edible protein content. He showed that in terms of protein production per acre, cabbages are a far more efficient crop than any animal, and by a very large factor: 1,100kg/ha of protein, compared with 115 for dairy produce, 102 for dairy and beef cattle, 27 for beef alone, 23 for sheep, 50 for pigs, 92 for broilers and 80 for eggs.

Soyabeans are one of the richest vegetable protein sources and although they do not grow well in temperate conditions, they are well adapted for tropical agriculture. Soyabean production is increasing. The 1970 harvest was 3 per cent higher than that in 1969, at 41,764,000 metric tons. However, the US produced 75 per cent of this total, and American soyabeans are processed into animal feed. Several of the developing countries, including Indonesia, Mexico and Colombia reported reduced yields.

Soya protein can be processed into food for direct consumption by humans and there can be no doubt that from a nutritional point of view this is far more satisfactory than incurring the losses that are inevitable if it has to be processed through animals. Dr. Magnus Pyke, of the Distillers Company Ltd., is a well known enthusiast for novel sources of protein. Soya protein, he has said, is marketed with some success as kosher bacon!

If once we accept such protein foods as those derived from soya-beans, the field widens, for protein can be synthesized. Some of the technical details of the processes involved are closely guarded trade secrets, but the principle is simple enough. A fungus – usually a yeast – is grown on a nutrient base in stainless steel vats, under strictly controlled conditions. At a particular point, the yeast is harvested, killed, and processed to extrude a substance that is rather like a man-made fibre. The fibre can be coloured and textured to give 'knitted steaks'. In theory it may be used to simulate any meat product. Synthetic protein of this kind is already in production and Shell, BP, ICI, Ranks-Hovis-McDougall and Courtaulds are among the major companies that have invested large sums in it. The USSR is reputed to have produced a million tons in 1970, for animal feeds. By converting non-edible organic matter into protein, yeasts act in a way similar to farm animals, but the level of efficiency is far higher, since the whole of the yeast can be eaten. The final product has a protein content higher than that of meat and, potentially at least, it is cheaper. The Russians are said to have used wood cellulose as their nutrient base. The most common base in use in Europe is petroleum or petroleum derived from coal. Dr. Pyke is hopeful about the possibilities. Since fats can be produced from petroleum or from coal, he has said, 'it only remains for the price to be right for margarine to become available from the National Coal Board!'

Fish-meal has also been used as a nutrient base; animal feed that yields its edible protein yet by-passes the animal, and on 5 May 1971, the New Zealand Minister for Science, Mr. Brian Talboys, announced that a pilot plant had been opened which recovered the protein from slaughter house wastes.

Considerable research and development has gone into the production of protein foods direct from green matter, including algae. Sometimes scientists have taken little known foods a stage further. The French Institute of Petroleum were seeking for an alga that could be used as a food additive and that could be grown in a climate similar to that in Chad. By the shores of Lake Chad they came across a tribe, the Kámenbu, that dries out and eats the algae that grow in the lake, and that has done so throughout its history. The tribe is 80,000 strong and doctors found not a single case of malnutrition. The alga, spirulina, is being grown now in Mexico and in Algeria. It should not be too unfamiliar to the Mexicans as a food source, because the Aztecs also ate algae.

If almost any plant can be processed into an edible and nutritious food, it is understandable that the scientists working in this area should be enthusiastic. It has been suggested that these synthetic or simulated foods offer a complete solution to the protein deficiency of much of the developing world. Professor Arnold Spicer, who leads the Ranks-Hovis-McDougall research team that developed one of these new foods, has said that the impact of them will be similar to that which man-made fibres made on the wool and cotton trades.

There are still problems to be overcome. Some workers have expressed concern that development is proceeding too fast and that the new proteins may not be free from harmful by-products acquired during the manufacturing process. There is difficulty, too, in persuading consumers to accept the new products. In spite of the relish with which they have been eaten by the scientists that produced them, there can be little doubt that they taste different from traditional foods. People are very conservative about the foods they will eat and most conservative of all are those living in more traditional and less sophisticated societies than those of Europe and America. Moreover, people adapt to the food that is available to them and to which they are accustomed. Not only may a new food be unacceptable to them, it may also be indigestible. It may not be recognized as food at all, as insects, frogs and snails are not generally recognized as food in Britain. So problems of palatability have not yet been overcome. It is psychologically important, too, that consumer-resistance be overcome in the developed countries first, otherwise people in developing countries will think, and with some justification, that they are being offered an inferior food that is not good enough for the rich countries.

Assuming that such difficulties can be surmounted, are synthetic proteins really a solution to the protein gap? The raw materials are, or can be, readily available. Or can they? Where synthetic proteins are produced from petroleum or coal, the availability of the raw material is very strictly limited and if the world were to commit itself to large-scale production from this source the best we could hope for would be competition for fossil fuels for use as sources of power or food. Yet algae can be grown without much difficulty, surely, and wood? Yes, they can, but world demand for timber already approaches the limit of productivity of the world's existing forests. There might be advantages in re-foresting what is now farm land, but there are physical limitations of space even then. Algae can be grown easily provided they can be fed – with fertilizers, for which they must compete with existing farm production for commodities that may become scarce in the near future. Protein may be recovered from wastes, provided there is an industry producing suitable wastes. In a finite system, such as obtains on this planet, there is no hope whatever of getting something for nothing. That would be perpetual motion.

All the same, the by-passing of the animal stage in protein production offers a greater efficiency and, therefore, more protein for the same input? This is true, but such is the complexity of introducing synthetic proteins for direct consumption by humans that much of the serious development at present aims to do no more than provide animal feed additives. Of course, this may lead to an increase in output of animal proteins, even though there will be a loss during the second conversion. This is probably the most useful application of synthetic protein.

Even for this more limited use, there are restricting factors. The product may be cheap if it is produced in large enough quantities, but the initial investment is high. It may prove too high for many developing countries unless aid is made available for this purpose. And the process consumes energy. Heat is required to coagulate the raw protein. The alternative to a heat process might be a chemical one, but that would consume even more mineral and energy resources. Also, the process requires water.

The countries that have the capital and resources to manufacture synthetic proteins are those that are rich already!

In the long run, then, synthetic proteins may provide a useful but limited supplement to human diets, either directly or as animal feeds. They will not 'take over' from 'natural' protein and their overall contribution to the world food problem is likely to be marginal.

We must beware too, in our enthusiasm, of attempting to simulate processes that may take place in nature with far greater efficiency. One of the more extravagant suggestions is that man should harvest marine plankton – krill – and process that. Since it is wild and, presumably, not difficult to obtain, it sounds as though here may be a food source that is available with the minimum of effort and that by by-passing the higher life-forms in the sea we may increase the protein output of the oceans. Apart from the energy limitations to which the process would be subject, it is interesting to compare the idea with the efficiency with which the blue whale converts krill to protein naturally. The blue whale mother converts 2 to 2.5 tons of krill a day into milk for her baby, which gains weight at the rate of 170 to 200lb a day. This is a conversion rate of less than 10 per cent, but could we improve on it when we take into account the energy that would be involved in powering ships and operating the processing plant? Perhaps it would be more profitable to husband the whales and harvest them in this case.

What, though, if we could obtain food from any growing plant? Let us assume, for a moment, that the production of synthetic protein from vegetable matter were practicable on such a scale that it could take over completely from any and every other form of food production. How many people would the planet then feed?

We must return to the most basic principles. The only constant, inexhaustible source of energy to power biological processes on this planet is solar radiation. This is made available to animals after having been trapped by green plants. The purpose of feeding is to transfer this energy to the animal in order to provide power for its metabolism. Man has inhabited this planet for something like a million years. He has farmed for 10,000 years. This is a long time in relation to recorded human history – to the history of our own culture – but it is only 1 per cent of the total. For 99 per cent of his time here man has found food without recourse to farming. Of the 80 thousand million human beings that have lived, more than 90 per cent have lived without any form of agriculture, 6 per cent have been farmers and the remainder have lived in cities. Man evolved into a natural system in which he took part, just like any other animal, in the utilization of the solar energy within the system. What was life like for him?

Our non-farming ancestors were hunter-gatherers and the stability and success of their way of life is evidenced not only by the fact that it survived for so long, but also by the fact that on a small scale it survives even today and only the most determined onslaughts by western culture will disrupt the life of hunter-gatherer tribes. There is a traditional view that life in a hunter-gatherer community was 'nasty, brutish and short.' The accounts of modern anthropologists who have lived among hunter-gatherers contradicts this view. Robert Waller* has summed up the more up-to-date picture of them:

'. . . the Kung Bushmen work on an average a 20 hour week. As hunting is unpredictable, they will sometimes work for a week and then rest for a month. The Hdzas of northern Tanzania spend most of this spare time gambling, which (in the dry season) takes priority over hunting.'

He goes on to compare their diet with that of their agricultural neighbours:

'It is far more diverse. Those who have lived among hunter-gatherer societies not yet invaded by other civilizations report that famine is unknown and that there are no signs of nutritional diseases. The Bushmen have eighty-five species of fruits, berries and melons and thirty species of roots and bulbs; fifty-four species of meat are classified as edible. Although the greater part of the regular diet is confined to twenty-three species of vegetable and seventeen species of meat, nevertheless, the rich variety available gives the Bushmen a wide range of alternatives and a good safety margin in time of potential want.

'Asked why he hadn't taken to agriculture a Bushman replied, "why should we plant, when there are so many mongongo nuts in the world?" '

Why indeed? The Bushman consumes an average of 56 grammes of protein per day from his mongongo nuts alone. This compares very favourably with the per caput daily protein consumption in all the countries of Africa. When all his protein sources are taken into account, the Bushman has one of the highest levels of protein consumption in the world!

The hunter-gatherer represents a biological ideal – man living in stability and perfect harmony with his natural environment. Of

* Out of the Garden of Eden, *New Scientist and Science Journal*, 2 September 1971.

course, he did not produce any Einsteins or Beethovens or anaesthetics, but then, he produced no Hitlers or Stalins or atom bombs either. Who can say whether the advantages of one way of life outweigh the disadvantages?

The discussion is unrealistic, of course. Agriculture made it possible for population density to increase, so that once begun its continuance became necessary to survival. Once the natural order had been disturbed and man took over the management, he had no choice but to continue or die. Waller suggests that this is a valid interpretation of the myth of the Garden of Eden and man's expulsion from it – agricultural man's, that is. Yet up to the time that agriculture began the average population density of humans over the planet as a whole was only 0.04 per sq. kilometre. Now it approaches 20.

We cannot return to hunting and gathering, but we can, perhaps, learn from the hunter-gatherers. Is farming the most efficient way to trap solar radiation and is our diet as diverse as it might be? The efficiency of the process must depend on the amount of photosynthesis that takes place and this in turn depends on the total area of green leaf that is exposed to the sunlight. The more green leaf there is, the more energy will be trapped. A considerable portion of the current research that aims to increase agricultural productivity is concerned with developing plants that will trap sunlight more efficiently. Until such crop plants are developed, some natural ecosystems are more efficient than our farms, and their increased efficiency can be measured in terms of their total production of dry matter. According to Professor P. J. Newbould*, bracken in Scotland will yield 26 metric tons a year from an area of land that will yield a maximum of 21 tons of wheat. Tropical rain forest, on the other hand, will yield 32.5 tons from a similar area. This is worth bearing in mind when desperate technologists suggest clearing the Amazon basin and farming it!

The success of the hunter-gatherer is due to the fact that he 'harvested' only the surplus production from the ecosystem. If, instead of selecting a few preferred plants and learning to cultivate them, he had sought to increase the productivity of the system as a whole, and

* Comparative production of ecosystems, *Potential Crop Production.*

continued to live off the surplus, might he have fared better than he did by becoming what we know as a farmer? Did the deterioration in his diet indicate that nature is better able to grow plants than man? Probably he had no choice but to develop in the way he did. Our position is different. Might we do better if we were to abandon farming as we know it today and to stimulate the productivity of natural systems? We are able to take more of the surplus than did our ancestors, since diverse though his diet was, we are able to process and eat foods that were not available to him.

Let us imagine, then, a world without farmers as such, a world which feeds us by producing the maximum amount of green matter possible. We have a large part of the earth's land surface at our disposal and perhaps we can add to this by the cultivation of algae, such as *Chlorella,* which is reputed to be far more efficient than agricultural plants at photosynthesis.

There is an ecological rule of thumb that says that the green plants in a natural ecosystem trap something like 0.1 per cent of the available solar radiation. Herbivores are able to consume some 10 per cent of this – the energy represented by the edible portions of plants – and carnivores have available to them 10 per cent of the energy that was available to the herbivores. From this it has been calculated that mankind has available to it about 1 per cent of the total solar radiation trapped by plants. Edward S. Deevey, Jr.,* has challenged the base figure – the efficiency of land-plant production – and suggests that it should be raised by a factor of three or four. Even then, his estimates may be low; Prof. Newbould's figures are higher than this for Austrian pine forest and for several agricultural plants! His estimate is in the region of 0.6 per cent. If he is right, and if we assume that the whole of this green matter can be processed into foods suitable for humans, then the present population of the earth could increase by up to sixty times – to something like 216,000 million.

Of course, there would be a price to pay. Man would have to become exclusively herbivorous. We would not be able to afford to feed at the reduced level of efficiency of the carnivore. Not only that, but man would have to replace all other herbivores. We could tolerate no competition for the food that we need – and our food

* The human population, *Scientific American,* September 1960.

now represents the entire plant production of the planet. There could be no other form of animal life of any kind. Nor could we allow land that will grow plants to be used for any other purpose. Land-based cities would have to be confined to the non-productive regions. Probably this would mean the polar icecaps and perhaps some of the most arid deserts, though urban development there would be restricted by the difficulty of supplying water (what there was would be needed for the plants). However, even the polar cities might not be large enough to accommodate all of us. In this case we would have to construct cities over the seas, built on piles, perhaps, or floated on pontoons of some kind. The oceanic cities would need to supplement food production, too, by surrounding themselves with submerged walls to make artificial lagoons in which mats of algal cultures could be grown, fertilized by human excreta.

Then, at this density, we would have reached the ultimate limit of human expansion. If it continued unchecked past this point, the whole edifice would collapse. If population continues to increase at its present rate, we will reach this population level in about 200 years!

What then? What do we do when we have reached this final limit? Do we try, somehow, to keep on growing? Do we seek ways to increase food production still further? How? Do we allow natural events to take their course and suffer a population collapse?

Or do we then, at the last moment of human history, seek ways of stopping growth? If this is a situation we are bound to face, should we wait to face it then, or should we not face it now? In point of fact, do we face it now anyway?

We have seen that the investment, in terms of capital and resources, that would be required to process green matter into protein in this way are colossal. Even if we were to abandon all other human activities, if we were to regard food production as the sole aim of our civilization, if we were to give up all our material possessions and live in the utmost simplicity, with the barest minimum of everything, could we afford, even then, to produce our food in this way and on this scale? Of course we could not. We would have no source of power with which to do it. The exhaustion of fossil fuels and the thermal and other pollution caused by the generation of energy on the surface of the planet mean that we must learn to harness solar radiation to provide all the power we require. Could we do this in addition to the effort that would be involved in turning over so much of our land to food production? Yet we would need, if anything, even more power than we consume at present, to process our foods and to maintain the life-support systems within the highly artificial environments within which we would have to live.

All these attempts to solve the world crisis in food are no more than evasions of the real problem. Most of them are unrealistic, if not wholly fanciful. True, some may contribute to the food we have available, but the relative size of their contribution is limited. The planet can feed only so many of us. Either we accept this and decide to limit our own numbers, or nature will do it for us.

11

What is Survival?

Survival is a word that figures prominently in much of the discussion of man's future. We have been talking about survival. What do we mean? When the final limit is reached in any one direction, what will life be like?

Let us imagine for a moment that the optimists are right. They are not right, as we shall see, but let us indulge them. Let us suppose, then, that the world really could support a human population of something like 200,000 million. How would we live?

Obviously there would be certain immediate problems that would have to be overcome, or to which we will, for the moment, assume solutions have been found. We would need power, for example. Let us imagine, then, that we have discovered some limitless source of power. Again, we would have to aim at a maximum housing density. This is because we would be prohibited from constructing dwellings on any of the land that could grow plants. The lack of water would rule out the desert, but desert or icecap, the environment would not be particularly hospitable. Therefore support systems would be needed and they would be needed on such a large scale that we would have no choice but to opt for economies of scale. It is simpler and cheaper to heat one large building than one hundred small ones.

It has been suggested, in all apparent seriousness, that a solution to the accommodation problem might be found in the construction of whole cities in the form of towers and that to achieve the maximum economy of ground area, these might rise as much as 40,000ft into the sky. Technologically, they say, this is feasible or soon will be. Such a tower would be divided into levels, rather like a block of flats, except that each level would contain an entire village, a complete community. Such a tower might contain a very large number of human beings. If we suppose that it is 40,000ft high, with a base diameter of one mile, and 200ft between each level and the next, there would be 200 floors, or villages, each occupying an area of about 1.5 sq. miles. The tower as a whole would provide a total floor surface area of 300 sq. miles. Populated at a density of 5,000 persons per sq. mile, the tower could contain 1.5 million people, all within a land area of 1 sq. mile.

Such a complex construction would have to contain its industries within itself. It is inconceivable that large numbers of people could leave to go to work outside. The working population of the entire city might amount to 500,000 people. If anything like this number were to attempt to leave or enter during a 'rush hour,' the congestion on the ground floor and around the exits would bring the city to a complete standstill on the first day. No, factories would have to be distributed inside, so that the very minimum number of people were required to leave.

This does not mean that the factories could be distributed evenly. It is doubtful whether this would be possible economically, at any rate within a 'free enterprise' system. Raw materials would have to enter at ground level and they could not be transported to the upper levels for processing because of the cost of lifting them, especially if they were heavy or bulky. So we might expect the ground floor to contain warehouses, so that it became a kind of dockland, close to the entrances. Factories would tend to distribute themselves with those engaged in the heaviest industries closest to the ground, so forming a heavy industrial area close to the docks. Lighter industries, part of whose supplies would include tools or part-finished products from the heavy industries, would be situated above them, while the upper layers would be given over to commercial, administrative and residential areas. The city would, in fact, arrange itself much like any other city of comparable size. The necessity for industry to locate itself close to the source of its materials would be no less, for although the distances over which materials and goods were to be conveyed would be much less, they would be vertical rather than horizontal, and since it takes more power to move objects vertically than it does to move them horizontally, the advantage of a small land area would cancel itself out.

The physical separation of the levels one from another would make it impossible for one area of the city to shade into the next as it might in a horizontal city. So the distinctions, of all kinds, between levels would be intensified. If the cost of the transport of industrial goods were to restrict their movement through the tower, this might be presumed to apply equally to building materials. The higher they were carried the more expensive they would become. So dwellings would increase in price with height. This would intensify still further the social stratification that would evolve.

Such stratification would be likely to produce a secondary effect. Already the inhabitants of the more expensive residential areas of our industrial cities and their suburbs are reluctant to visit the industrial and dock areas except on business. If the degree of stratification were to increase, this tendency might increase also. However, in the tower city, it would be necessary to pass through the warehouse and dockland area in order to reach any exit. This might discourage the inhabitants from the upper levels from leaving the city at all. It is well known that in tower blocks of flats there is a reluctance on the part of some of the people living on the upper floors to go out of the building, simply because of the effort involved in moving considerable distances vertically and also because of the feeling of isolation from life on the surface that is engendered by life at a high level. In the tower city the effort involved in leaving the building would be much greater than that required to cross the centre of an industrial horizontal city and to leave through the dockland. In a fairly short space of time after the people moved in, we may anticipate that many would cease to go out at all.

So we might expect such a tower to become not only self-supporting, which it would have to be, but almost completely sealed from the world outside. Once moved in, the population would remain inside for ever.

They would lose touch altogether with the outside world. A tower rising to such a height could have no windows. This would be structurally impossible because of the thickness of the walls. Air pressure would have to be maintained artificially and oxygen would have to be circulated. For this reason, the doors through the walls at the base of the tower would need to be sealed hermetically. Entry and exit would be through air locks, for if the doors were allowed to remain open, air would escape, so reducing pressure on the upper levels. The isolation from the outside world would be virtually complete.

Without windows, all lighting would be artificial, day and night simulated and produced by the City Engineer's Department.

The city would be a technological masterpiece, but the whole of its essential life support systems would be dependent on machines. A failure of power would bring chaos to any major city today. To the vertical city it would bring death. A failure of the air locks might cause anoxia on the upper levels – indeed, depending slightly on individual susceptibility, to the whole of the upper half of the tower. A failure of the air conditioning might lead to a rise in temperature, again, especially on the upper levels. There would be heat exhaustion, accompanied by a build-up in the carbon dioxide levels lower down. A lighting failure would cause chaos at any time of day, not just during the evening or early morning. In many cases, a system's failure would affect the upper levels most quickly and it would be of political significance that those would be the levels containing the most desirable and expensive residences.

Food would be rationed at barest subsistence level: we are considering life for the maximum possible human population, after all. However, our knowledge of existing societies suggests that no matter what legislation were passed, no matter what sanctions were applied, a black market would prevent the equitable distribution of the food that was available. The rich would eat better than the poor. Since the basic ration is that required for survival, and no more, sections of the population would be starving or suffering from malnutrition all the time. These people would live on the cheaper, lower levels. The sections of the community would be polarized still further.

So this would be life in the most densely packed city conceivable. The division between town and country would be absolute: there would be no countryside anyway. The city-dwellers would never leave the city. They would never see natural sunlight. Most of them would be hungry. The richer of them in particular, but eventually all of them, would be dependent for their lives on the efficient functioning of machines and the loyalty of those with access to the machines – one act of sabotage could kill hundreds of thousands.

Is this the brave new world? Is this the life we would wish for our children and their children? It may be what we are planning.

It would not be a healthy life. Based on a barely adequate diet of synthesized food, nutritional diseases would be endemic. Even today in the most affluent societies, diseases of malnutrition, including appendicitis, obesity, dental caries, peptic ulcer, coronary thrombosis and a host of others, are increasing. In our vertical city these might be the illnesses of the upper levels. Lower down they would be joined by the more traditional ones – scurvy, beriberi, kwashiorkor. It is likely that the incidence of cancer, increasing fast today, would rocket. Added to these would be the risk of infectious disease. The sickness services – it would be an absurd euphemism to call them health services – would need to be larger and more efficiently organized than they are today if they were to hope to control this human battery house.

What of the psychological and behavioural effects of such pressures as would exist? There is data today that suggests definite causal relationships between population density and violent crime. In the August 1970, issue of *The Ecologist*, Claire and W. M. S. Russell described the behavioural effects of overcrowding on zoo monkeys:

'In monkeys living under completely relaxed conditions in the wild, quarrelling is always rare and violence is practically nonexistent. In monkeys living under the stresses of an enclosure, quarrelling is frequent and violence appreciable, sometimes leading to serious wounds and death. This contrast applies to all the species studied ... the zoo monkeys are clearly getting an ample food supply, and can be regarded, so to speak, as materially "affluent" societies. What they lack is *space* ... In a colony of hamadryas baboons at Regents' Park, observed in the 1920s by Sir Solly Zuckerman, eight out of sixty-one males died by violence, but thirty out of thirty-three females and five out of five babies were killed in this way.'

The Russells are hesitant, rightly, about extrapolating from monkey behaviour to human behaviour. '... while all mammals are capable of fiendish cruelty under stress, man differs from them all in having an even more highly developed positive parental urge.... Even under frightful stress, where almost all mammals would kill their young, many human beings continue to love and protect them at great sacrifice to themselves.' But what is overcrowding? Man has almost always lived at a density that has pressed on his resources, and so most of his history has been spent under some degree of stress from overcrowding. Yet, in his *Autobiography,* H. G. Wells sounded and ominous warning – and he was describing an episode in his early life:

'I looked, so to speak, through a hole in my life of some weeks more or less, into a sort of humanity coarser, beastlier and baser than anything I had ever known before ... I think that the peculiar unpleasantness of that episode lies in the fact that we were all too close together. We were as congested as zoo monkeys.'

We live in an age of violence. Crimes of violence increase constantly and sizeable areas of some of the world's major cities are rendered uninhabitable because their citizens cannot be afforded adequate protection by authorities equally violent. Our wars tend to increase in size and scale. There is little reason to believe they decrease in frequency.

Yet today, except in a very few cases, no city exists cut off from a hinterland into which its population can escape for some kind of relaxation, if not complete solitude. If we compare the situation in the cities of the world today and those within our infinitely more claustrophobic vertical city, we must conclude that the tower could not long survive. Crime would lead to repression, which would increase the violence. There would be civil war and riot, but with the difference that in this case the first fanatic to smash the life-support control machinery would condemn all of his fellow citizens to death. Would anyone be so insane? Almost certainly they would, for the first to be affected would be the rich, the privileged who lived at the higher levels. The poor could escape through the exits into the world outside: into the arctic wastes.

So the tower city could not endure. Could it be built at all? If we leave aside the constructional problems, which might be considerable (!) we are still left to locate a source of power. This deserves a little more attention, for the provision of cheap power is a basic requirement for the continuation of our present industrial society and it is necessary for the development of industry and, indeed, of food production, in the developing countries.

At present our main sources of power are the fossil fuels – coal, petroleum and natural gas. As we saw earlier, the world reserves of petroleum and natural gas will be exhausted by the end of the century and if we revert exclusively to coal, this will not last long either. The popular view is that nuclear power will provide an inexhaustible alternative. Unfortunately, this is not true. At the present time the development of Britain's atomic power programme is bedevilled by difficulties. It is proving far more expensive than had been anticipated – expensive for a rich, developed country, not a poor one – and it seems unlikely that the cost of power produced in this way will fall. This explodes the hope for cheap power. In spite of the many rigid safeguards imposed on the nuclear reactor process and on the disposal of its radioactive waste products, accidents occur. In one widely publicized incident at the Windscale plant in Cumberland, more radioactive materials were released into the atmosphere than were released at Hiroshima. It was fortunate for the British people that the wind carried them out to sea, not inland. The wastes that are stored must be kept intact for centuries and the pursuance of this policy is possible only on the assumption that our civilization will continue for many hundreds of years to come, or that, should it fall, it will be replaced by one whose people understand that these stores must not be tampered with. Is this a responsible assumption? Apart from accidents, environmental pollution from leakage is serious today and it increases with every new plant that opens. It has been estimated that if the proposed programme were to proceed according to plan, so that by the end of the century we were to derive most of our power from this source, radiation levels would exceed the internationally accepted maxima. There is no safe level of exposure to radiation. 'Safe' levels are based on what is deemed to be a socially acceptable disadvantage in relation to the social advantage derived from the power generated. The Society for Social Responsibility in Science, in America, which is composed of experienced and informed scientists, has recommended that no further licences be granted for the operation of nuclear reactors until the pollution problems have been solved.

Even if the present reactors were to be made so safe and cheap that they could be allowed to increase in number, their fuel, uranium, will be exhausted before the end of the century at present rates of consumption.

Unless a new source of fuel is discovered, it is generally agreed that the nuclear power phase will be short-lived. The solution most usually proposed is the fast breeder-reactor. This generates power, but it also produces fuel for itself. Theoretically it is ideal, for its net fuel consumption is far lower than that of a conventional reactor. Britain plans to convert to fast reactors as quickly as possible and its first, at Dounreay in Caithness, was opened in September 1971. At that time there were seven fast reactors in the world that were operational, and three of them had experienced major accidents involving the release of radioactive substances.

The third nuclear alternative is the fusion reactor. This would be based on the fusion of hydrogen atomic nuclei, rather than the fission of uranium – the principle of the hydrogen bomb, rather than of the atomic bomb. Its main fuel would be water, thus making it the most attractive proposition of all. So far the fusion reactor is little more than an idea in the heads of certain physicists.

Industrial society could not survive a hiatus, a period without a cheap power supply. If it is to survive in its present form, therefore,

Radioactive wastes as a Function of expanding US Nuclear power

	YEAR *1970*	*1980*	*2000*
Installed nuclear capacity MW(e)	11,000	95,000	734,000
Volume high-level liquid waste[a,b]			
Annual production, galls/yr	23,000	510,000	3,400,000
Accumulated vol. galls/yr[c]	45,000	2,400,000	39,000,000
Accumulated fission products, megacuries[b]			
Sr^{90}	15	750	10,800
Kr^{85}	1.2	90	1,160
H^3	0.04	3	36
Total for all fission products	1,200	44,000	860.000
Accumulated fission products, tons	16	388	5,350

(*Blueprint for Survival*)

[a] Based on 100 gal of high-level acid waste per 10,000 thermal megawatt days (MWd) irradiation
[b] Assumes 3 yr lag between dates of power generation and waste production
[c] Assumes wastes all accumulated as liquids.

an alternative fuel to petroleum must be fully operational in time to make a phased take-over.

There is a further limitation. The largest single source of energy is solar radiation and it is this that maintains the planet's surface temperature within certain limits. The consumption or combustion of other energy sources releases heat and all energy generated on the planet ends as low-level waste heat. This tends to raise the surface temperature. While this presents local problems at present, it does not affect the planet as a whole because the amount of energy generated by man is infinitesimal compared with that derived from the sun. However, if we were to continue to increase our energy generation and consumption at the rate that is necessary for continued industrial growth and for the industrialization of the developing countries, within the next hundred years man will be releasing into the environment sufficient heat to begin to raise the temperature.

The only solution to this problem and, in the end, the only permanent source of energy, is the sun itself. If we were to increase the efficiency with which we utilize solar radiation this would not produce global thermal pollution problems, because we would not be adding to the amount of energy at the planet's surface. The solar radiation is there, whether we use it or not. At present, we are a long way from harnessing solar energy to industry.

Let us return to the tower for, fanciful though it may seem – and I emphasize again that it is counted as a practical solution to overcrowding by some people – most of the arguments both for and against it might apply equally to any other plan to achieve a maximum population density. The most likely sites for such cities would be the icecaps, because of the easy availability of fresh water, but the alternatives – the deserts or the seas – are equally harsh and thus large, communal buildings would be essential in order to achieve the maximum conservation of resources and maximum protection against the climate. So the tower seems less fantastic. Even then, if we are to accommodate 200,000 million people, we would need to build something like 135,000 cities the size of our tower. Would such a large population living at the poles melt the icecaps?

In reality, of course, such proposals are little less absurd than those to export surplus people into space. Such cities could not be built with present-day technology, they could not be supplied with power, and the overcrowding within them, combined with the artificiality of the internal environment, would produce social stresses that would cause violence to erupt within a very short time.

The exercise is worthwhile, nevertheless, because it shows that in the end there are other limitations to human expansion than food. Even if the planet could feed 200,000 million people, it could not house them nor supply them with the other necessities of life and it could not prevent them from killing one another. Man does not live by bread alone.

If not 200,000 million then, how many? There is no single answer to that once we abandon the concept of bare survival. Survival is not enough. Even then, there are many people to whom it appears attractive, conjuring for them a vision of a pleasant, ascetic life. There is the minimum of food, but it is always good, wholesome, palatable food. There is a minimum of living space, but always there is access to areas of romantic wilderness. There is a minimum of material comforts, but always there is access to the middle class delights of music, art and stimulating conversation. Survival for the maximum number would produce intolerable conditions, as we have seen. Survival at a personal level, too, is far less attractive than it may sound.

William Golding has explored this kind of personal survival in his novel *Pincher Martin.* Pincher Martin was drowning in the sea and he prayed to be allowed to live. All he wanted, he believed, was his life, nothing more. His prayer was answered – literally. He found himself washed up on a small rock in the North Atlantic. There were depressions in the rock surface in which rainwater collected, which he drank. There was a variety – a small variety – of shellfish, which he ate raw. There was a hole in the rock into which he could fit half of his body to protect it from the worst of the weather. That was his survival. When we demand life at any price, we may find the price is too high.

No, we must consider the quality of life. Deevey, after calculating the theoretical upper limit for population in terms of food supply, goes on to consider the problems of psychic stress, referring to experimental work among rats, where it was found that biological controls came into operation and numbers came to be controlled through the automatic suppression of spermatogenesis and ovulation and through foetal resorption and then infanticide. He concludes:

'I am not implying that the population explosion will be contained by cannibalism or foetal resorption, or any power so naked. I simply suggest that vertebrates have that power, whether they want it or not, as part of the benefit – and the price – of being vertebrates. And if the human method of adjusting numbers to resources fails to work in the next 1,000 years as it has in the last million, sub-human methods are ready to take over.'

But how many? Perhaps we should consider how many people could survive with a standard of living comparable with that of Western Europeans or North Americans. This, after all, is the life we have been promising the inhabitants of the developing countries for the last twenty years.

We saw earlier that within the next two decades we must discover and make operational an alternative source of energy to power our industries. For the immediate future the most likely candidate is coal, provided a stable, renewable substitute is being prepared to take over when coal supplies begin to run short and when coal is required for its by-products rather than for combustion. It is probable that there will have to be economies, too, in our use of non-renewable raw materials. Minerals are being mined out and many of the substitutes – particularly plastics – are dependent on the petrochemical industry for supplies or organic chemicals. We cannot count on them. We will have to learn to build to endure and to recycle all our wastes – no matter what the cost. If the present economic systems of the world prevail, we will have to become self-sufficient in all our requirements, for just as the domestic demand for food in the developing countries will mean severe reductions in the volume of international trade in foodstuffs, so the consequent retarding of economic development in those countries will make it more and more difficult for them to buy manufactured goods from us. It is certain, therefore, that most of the countries that are industrialized today are, in fact, over-developed. If they are to survive they must find ways to 'de-develop.' This will mean a reduction in their material living standards, an abandonment of the dogma of economic growth. In such a situation it is probable that Britain could support a population of no more than 20 millions.

Yet we find ourselves discussing the possibility of the developed countries maintaining their present living standards, rather than the possibility of the countries of the third world benefiting. Obviously, none of them can possibly become industrialized in the way that Europe and North America are industrialized.

This non-answer has answered one question, however. We know now that no country can sustain the present level of prosperity – of production and consumption – of the developed countries, not even those countries themselves. We see, too, that the developed countries are already grossly over-populated and that if they were thrown back on to their own resources this fact would immediately be apparent.

Let us then lower our standards and see whether we can continue at, say, the level of Southern Europe. For Europeans and Americans this seems more likely. There would be a limited amount of industry, consuming a limited amount of power, and the regions would produce all their own food. The nature of the industry might change, but the scale is about right. Yet even then we can reach no satisfactory conclusion because the most basic requirements of people vary with climate, with topography, with culture and with tradition. If we say that a living standard equivalent to that in Greece, say, is sustainable throughout Europe, we must take account of the fact that Scotland is colder than Greece. We may find that Greece can support a denser population than Scotland, or Kent, or Connecticut.

What is apparent is that our figure of 200,000 million is falling rapidly. We are discussing not by how many present populations may increase, but by how many they need to be reduced. So perhaps we should consider whether the world can long sustain the population it has at present.

We should start by assuming that the available wealth and food were distributed evenly, but is this a realistic assumption? As things are, they are distributed very far from evenly. All the same, it is difficult to calculate how many people there might be when a tiny minority with all the power insists on taking for itself far more than its fair share. This lion's portion is not distributed equitably even within the rich countries. Europe and America have a very small number of very rich citizens, a rather larger number of – by third world standards – wealthy citizens, a larger number still of citizens living on a level somewhat comparable with that in a developing country and a minority that is left to starve. The 'western way of life' is enjoyed by very few people, so that the promise held out to the developing countries is of a living standard that is largely non-existent. Moreover, the developing countries already have wealthy citizens of their own; they are one of the obstacles to progress! So what are we offering? What is the 'western way of life?' Is it immortality in a Hilton hotel, or is it a few disease-racked years as a redundant Welsh miner?

Without major global political and economic reforms, the world is over-populated now. When more than half of the existing population cannot be fed or housed or clothed, when cities, that should be the cultural centres of our urban civilization become uninhabitable because of crime, pollution and congestion, when nations and sections of nations go to war over the remaining resources, how can we talk of a maximum population? We have exceeded the maximum already.

We cannot afford to gaze into the future in order to imagine a Utopian solution to the problem. It is here and it is now and it is us. We have about twenty years. If industry is to continue, it must continue with the resources and power it can find now, because it will take twenty years to make any modifications that are required. If the search for alternatives fails, if the modifications are not made, then industry will not continue. It is as simple and as final as that.

If the world is to be fed, it must be fed with the resources of land, water, sunlight, chemicals, equipment and workers that it has now. If it is not fed in these terms it will not be fed.

Professor Ehrlich has pointed out that there is no natural law that says that things are bound to work out all right in the end. There is no reason to suppose they will, unless we make them do so. Still, we have a touching faith in our technology. We persist, in the face of all of the evidence, in believing that it can save us from catastrophe with no effort from us, a catastrophe for which technology itself is largely responsible. David R. Brower, of Friends of the Earth, a voluntary organization dedicated to involving individuals in direct action to avert ecological disaster, has put our technology in the context of a longer time scale:

'Place the six days of creation in Genesis against the four billion (thousand million) years or so of the Earth's age. That is a ratio of 8,000 years to one second. All day Monday and up to Tuesday noon is the sheer formation of matter until there is a globe with oceans on it, and mountains. Then a cell begins to undergo mitosis, dividing in two, and life is aboard the planet by Tuesday noon. All the rest of Tuesday and Wednesday, Thursday and Friday, and well into Saturday, life expands and becomes more diverse, more stable, more beautiful. At four o'clock on the afternoon of Saturday, the last day of creation, the age of reptiles comes on-stage; at nine o'clock, it goes off stage. Just before the age of reptiles ends, there are redwoods – and just before redwoods, the pelican, a 90 million-year-old life form now threatened with extinction by DDT and man's urge to usurp the Earth. At three minutes before midnight, man appears. One quarter of a second before midnight, a bearded man, anti-establishment, talking of peace and brotherhood and Christianty is on the planet.

'Then, one-fortieth of a second before midnight, enters the industrial revolution. You have heard it predicted that we can go on at this rate that has worked so splendidly for one-fortieth of a second because "technology knows all the answers." I am urging that we worry just a little bit about how much we do know.'

This, then, is survival. We must consider, not whether the human population can expand, or by how much. Our concern, and it is the only concern of any relevance at all in the world today, is for how long the planet can sustain, even at its present inadequate level, its present population, and to what level the population must be reduced before stability can be achieved. We must solve this one. Should we fail, the next question will be, can man survive at all? If we aim to succeed, on the other hand, we will need to know what immediate steps we can take to gain time.

12

Waste not, Want not . . .

Food production and food availability are not the same thing. In India, in one year, the wastage of food between harvest and its reaching the consumer accounted for what amounted to 20.5 grammes of protein per head of population. In 1964, the wastage of foodgrain after harvest in India amounted to 47.3 per cent of the total crop. Similar figures can be produced for many other developing countries. In Nigeria, for example, 46 per cent of sorghum is lost and 41 per cent of the cowpea harvest. In Sierra Leone, 41 per cent of the rice crop and 14 per cent of the maize is lost. In tropical Africa as a whole, the total wastage after harvest accounts for 30 per cent of all the crops grown.

Clearly, in the developing countries, a reduction in this waste could have an immediate and profound effect on food availability and therefore on the standard of diet throughout large regions. The FAO is well aware of this, and has declared a 'War on Waste.' Its definition of waste is wide, but acceptable. According to the FAO, waste is 'avoidable loss.' Considering the size of the problem, however, it is surprising that the Organization does not pay more attention to it. Its *Speaker's Guide*, headed 'War on Waste,' begins resoundingly:

'Tens of millions of people who are hungry and ill-fed could be nourished with the agricultural products that the world now wastes.

'In the developing countries, 80 per cent of farmers are subsistence farmers with only just enough over from good years to keep them going through lean years. But this is often because much is wasted. Otherwise they would have a surplus over for sale.

'Waste is *avoidable loss* that means the world is squandering its resources in a reckless fashion because of lack of technical knowledge or its application.'

The leaflet then goes on to outline the problems associated with bad farming practices: crop loss due to pests, weeds and disease, faulty irrigation, pollution, bad distribution and the non-utilization of by-products.

Mr. A. H. Boerma, FAO Director-General, mentioned this lack of attention in a brief passage in the speech he made at the closing ceremony of the Second World Food Congress:

'The third Area of Concentration is our war on waste. It is not, I think, often realized that billions of dollars worth of food are lost or wasted every year – due to pests and diseases, to human negligence, to misuse of natural resources, and so on. This, frankly, was one area which I feel did not receive enough attention at the Congress.'

The problem is not confined to the developing countries: the United States loses something like 20,000 million dollars worth of food each year through waste. Much of this waste occurs through crop pests and diseases and weeds and the solution to those problems probably lies in improving the overall standard of husbandry. It is significant, perhaps, that the country with the most sophisticated chemicals for crop protection is nevertheless faced with such staggering and seemingly irreducible losses year after year.

Losses of grain after harvesting are due mainly to bacteria, insects, moulds and rodents. These losses could be reduced greatly by the improvement of storage and drying facilities. An increased proportion of aid spent on providing sealed grain bins, made from cheap plastics and easily transported to remote areas, might bring larger returns than the same funds spent on increasing production – to feed still more pests. In some parts of the third world, in India especially, people have religious objections to the taking of animal life. They will not kill pests. This scruple is deep-rooted in Hindu and Buddhist culture, it has a strong ethical base and quite apart from the almost insuperable difficulty of changing fundamental attitudes developed over many centuries and affecting millions upon millions of individuals, it is undesirable that it should be changed. Nevertheless, the exclusion of pests through improved storage should prove acceptable, even though this would imply their eventual death from starvation! It has been discovered that a four inch high parapet all around a grain store will keep out rats and mice; they refuse to cross it. It is possible to treat grain so as to repel rats. The technique is to pack the grain in treated bags and to stack these under polythene sheeting. The stack is then fumigated with either a mixture of ethylene dibromide and methyl bromide, a phosphine ammonia blend or an ethylene oxide formulation, depending on the type of infestation.

Improved drying may involve the burning of scarce fuel, but it may mean no more than exposure to the sun in protected areas. It could reduce losses due to bacteria and fungi.

Control of insects is more difficult, for apart from the other problems associated with the use of pesticides, it is no simple matter to develop one that can be used on food in store. There is no opportunity for the insecticide to disperse in the store, the time spent by the food in store being unpredictable it cannot be guaranteed that the insecticide will deteriorate before the food is consumed – and if it were to do so too soon it would lose its effectiveness, and a substance that will poison an insect may also poison a human being. Insecticides generally are used to deal with attacks at particular stages in the growth of a crop plant. If the plant can survive the attack it may proceed to the next stage. Thus, in theory at any rate, the insecticide may be allowed to deteriorate and disappear once the vulnerable stage has passed. Obviously this does not apply when the stage during which the plant is to be protected is that immediately preceding consumption.

Even then, however, the problem may not be altogether insoluble. The Central Food Technological Institute at Mysore, India, has developed a range of products that are non-toxic to human consumers. The first consists of a fine abrasive that reduces the thickness of the insect's chitinous outer skin as it crawls about. Insects depend for their survival on maintaining a certain level of internal water and this level is higher than that of dry grain. The weakening of the skin allows water to escape into the relatively drier conditions that surround the insect, and it dies from dehydration. This effect can be enhanced by adding an otherwise inert and innocuous substance – a clay, for example – to absorb moisture, and otherwise harmless growth promoters, such as sugar or tricalcium phosphate. These cause the larvae to make excessive demands on their bodily food reserves and they are left with insufficient to pupate.

The problems of improved storage and drying facilities do receive some attention. In 1970, the Intermediate Technology Development Group Ltd., a London-based organization devoted to raising productivity by making the maximum use of labour, skill and capital in developing areas, formed a Food Technology Unit, which explored the possibilities of improved rice storage in Liberia. The main part of the unit's work is concerned with developing and introducing simple food-processing techniques which help to make better use of foods. The project is run in collaboration with the National College of Food Technology, Weybridge, Surrey, and the funds for the Liberian project were provided by Christian Aid.

Christian Aid has also provided a grant of £53,240 for a somewhat similar project in Ethiopia. There, the work is being undertaken by the UK Committee of the FAO's Freedom from Hunger Campaign and assistance is being given by the Overseas Development Administration of the Foreign and Commonwealth Office and the British Embassy in Addis Ababa. The government of Ethiopia is providing counterpart officials to work alongside the British scientists. The extent to which governments are now aware of the size of the waste problem is illustrated by the fact that this project follows the recognition by the Ethiopian government of the need to train staff and farmers in improved storage methods, and that this recognition on the part of the government was the result of a survey by the British Tropical Products Institute, backed by the UK, which revealed the seriousness of the losses of maize, sorghum, rice and beans. The Ethiopians had not known just how much they were losing.

It would be naive to underrate the difficuty of changing the habits of generations of farmers and merchants, but the need to do so is acute. The FAO estimates that throughout the world as a whole, more than 10 per cent of the food grown never reaches the human consumer. 'Over 55 million Africans could be fed in a year on the grain eaten by rats, locusts, beetles, moths, weevils, quealea birds and countless micro-organisms.' If half of the 40 per cent of grain lost each year in India could be saved, food availability in that country would increase by one-third, with no increase in production.

A reduction in this wastage would bring other benefits. Rats transmit 35 different diseases to man and animals and carry 18 different kinds of fleas, lice and ticks. Eczema, mange, asthma and a number of allergies are caused by the larvae of grain pests and there are respiratory ailments caused by fungal spores.

In part, the extent of wastage must depend on the length of time spent in storage. In countries where hunger is commonplace it would be natural to assume that this length of time is sufficient only to transport the food to the market. Unfortunately, in the case of food that can be stored for prolonged periods, such as grain, this is seldom the case. It is more profitable to hoard grain bought at harvest time, when the supply is relatively plentiful and the price low, to resell later when shortages have re-appeared and the price has risen. The solution to this aspect of the problem calls for more than technological inputs, or education. It is a matter of political, social and economic reform. It is a part, a small part in fact, of the much larger wastage of resources and people caused by inefficient political structures. It is part of a complex of difficulties associated with ensuring that the benefits of improved production, and the means to achieve it, reach those sections of the community for which they are intended, so increasing food availability at the same time as productivity.

In India it is often the rich farmer who grows the new cereal varieties. It is only the rich farmer who can afford to buy the seed and then the fertilizer it requires. The increased yield will call for additional labour at harvest time – this is one of the aims – but the modern, progressive farmer may prefer to go the whole way and introduce machinery. This replaces workers and adds to unemployment at the same time as it adds to the profit of the farmer. According to Wolf Ladejinsky, of the International Bank for Reconstruction and Development, the new technology may be displacing agricultural workers, rather than providing more employment for them. This leads to an overcrowded labour market and low wages. The benefits are not reaching the poorer sections of the community.

However, this is not all. As the productivity of the richer farmers rises, land values increase. This leads to increases in the rents of tenant farmers, themselves too poor to grow the new varieties, and to an increase – from the traditional 50-50 to as high as 70-30 – in the landowner's charge to share-croppers. Tenants and share-croppers – the poor – are forced out of agriculture altogether, or are reduced to the status of hired labourers. This, too, works to the advantage of the landowner, because of the high value of his land, worth still more if there are no tenants. This sequence of events produces a polarization between rich and poor, and this in a country where 38.52 per cent – 160 to 200 million people – of the rural population lives below the poverty line.

As in many countries, India needs reform of an antiquated and inadequate system of land tenure. Reform is difficult and, perhaps because in poor countries the rich have so much power and influence, governments tend to be reluctant to attempt it. Of India, Wolf Ladejinsky wrote in the *Financial Times* (25 January 1971):

'All Indian States have enacted reform legislation, with such principal provisions as limitation on holdings (land ceilings) to form a pool of land for redistribution among tenants; security of tenure for those who remain on the land as tenants; and reduced rentals.

'As between anticipation and fulfilment, only a relatively small acreage has been redistributed and some tenants were made more secure on the land. On balance, however, most enactments were shot through with loopholes and, even where not, the implementation was at best half-hearted and at worst a signal failure. It is not unreasonable to speculate whether, as a result of massive tenant evictions in the guise of "voluntary surrenders" following the enactments, more tenants lost than gained from the reform movement.'

Ladejinsky goes on to point out that the situation is not completely black. Mrs. Gandhi has made persistent efforts for two years to persuade the chief ministers of the States to implement their legislation, and although this has been relatively ineffective so far, there is a fear that the 'Green Revolution' may turn red! History provides many instances of reforms that were introduced and made effective because the alternative to the loss of some power, wealth and prestige by the élite was the loss of all of it.

It was Miss Mercedes Concepción, moderator of the plenary session of the Second World Food Congress that discussed population growth in relation to agricultural development, who warned that the success of the 'Green Revolution' might do no more than substitute an employment problem for one of hunger.

Land reform is essential in very many countries, if any programme of agricultural development is to be successful. In Chile, for example, 40.8 per cent of all agricultural families are peasants, yet between them they own only 7.4 per cent of the farm land. From this small parcel, often of inferior land, the peasants produce 20 per cent of Chile's total agricultural output. Similarly in Brazil, 23.5 per cent of the agricultural population is made up of peasants, who own 6.5 per cent of the land. The Brazilian peasants produce 21.3 per cent of Brazil's total agricultural output. There are reasons for this relatively high productivity. The peasant, with a very small holding, must produce the maximum if he and his family are to survive. He has the labour of his wife and children to assist him and his land is cultivated very intensively. The large landowner, on the other hand, is under less pressure to maximise output and the labour he uses is employed labour, which must be paid. Thus the large estates generally are farmed less intensively than the peasant holdings. In a way it is in the interest of the landowner to farm the way he does. Of course, this is not to write off the landowner altogether. He may be able to engage in forms of husbandry that may be of great importance to his country and to the region as a whole and which would be impossible for the peasant. In Latin America, the production of beef is of considerable economic importance. In 1969 (preliminary figures) the region produced a total of 9.64 million metric tons of meat. This is not large compared to Western Europe (18.11 million metric tons) or North America (23.54 million metric tons), but it is far higher than the meat production of any other developing region and in Argentina there is ample protein for everyone – 87.7 grammes per caput per day, of which 50.7 is animal protein. The equivalent figures for the UK are only 88.0 and 54.0 grammes. Animal husbandry for meat requires either large areas of land for extensive grazing or high capital investment coupled with intensive arable farming for intensive, indoor rearing. The Latin American peasant has neither of these requirements, but the estate owners have land at least. Beef ranching and sheep grazing are practical and profitable propositions.

In terms of total food production, however, there can be no doubt that peasant farming is more efficient. It produces food more intensively and provides more employment, so ensuring the equitable distribution of the food grown. If overall production is to increase, then it would be advantageous to extend peasant farming and there is room for such extension, for in Chile 49.7 per cent of all agricultural families are landless and in Brazil, 61.9 per cent.

Alan Riding wrote about the situation in Mexico in the *Financial Times* (14 August 1971):

'Agrarian reform has long been a pillar principle of the Mexican revolution and land continues to be distributed to the peasants as a political gesture by successive governments. But even if the peasants in their Ejido co-operatives were equipped to farm properly, the land they receive is often poor – only 15 per cent of the country is suitable for cultivation and much of this is still in private hands.

'The social implications have been disastrous. The absence of work and hope in the countryside is driving hundreds of thousands of peasants to trek to the cities each year, adding unemployment and slums to the problems of the main urban centres. Yet even with this "valve," the number of farm workers is expected to rise from 5 million today to 8 million by 1980.

'Thus, if Mexico is ever to become self-sufficient in food and export enough to counterbalance continuing imports of manufactured products, a total reform of the agricultural sector will be needed.'

Mexico's population is increasing at the rate of 3.4 per cent per year. In 1967, the programme of reform had increased the proportion of landowners from 3 to 50 per cent of the rural population, but nearly 85 per cent of the holdings, occupying 74 per cent of the land, were less than 12½ acres in area.

In Chile, where a social democratic government came to power in November 1970, the political left is taking matters into its own hands to speed agricultural reform. Although he claims to be a Marxist, President Allende's policies are moderate, and the coalition government which he heads is reluctant to take radical steps to speed up development, or reform. Ministers are reluctant to repeat what they regard as the mistakes made in Cuba. Yet there is a political organization, the Movimiento Izquierda Revolucionaria, composed mainly of middle class students, that is radical. President Allende is troubled by the MIR, but in the event of problems with the right wing he would need MIR support. MIR has obtained funds by raiding banks and arms by stealing them. In rural areas it has speeded agrarian reform by seizing farms and turning them over to the peasants. Up to the November election, the MIR is said to have taken over some 300 farms and its activities have continued since. The situation is critical, since the government chooses not to interfere with these take-overs. Critical it may be, but reform is vital.

The whole subject of agrarian reform was discussed at the Second World Food Congress and one of the eight commissions was concerned exclusively with 'People in rural development.' After its last session, the commission submitted its recommendations to a plenary session of the Congress. It suggested a strengthening of existing organizations concerned with rural reform and advised that the FAO set up an advisory body to survey all FAO projects and programmes to ascertain their 'social and institutional feasibility.'

The recommendations were rejected by the plenary session. Several speakers said they were too vague. A Latin American said the document produced by the commission 'amounted to nothing' and would be ignored by totalitarian governments. A Swedish delegate said it was full of 'innocuous platitudes.'

There was a large number of young people at the Congress and it was one of them, Manuel G. Arejola from the Philippines, who accused the commission of producing a document that lacked courage and was not relevant to the needs of his country.

The young people were accommodated in an army camp on the outskirts of The Hague, renamed the New Earth Village. There they held meetings of their own and principal speakers from the Congress, including Mr. Boerma, visited and talked with them. The New Earth Villagers produced a paper that aimed to strengthen the recommendations of the commission. The paper said:

'We realize that as long as there is no world government or any other higher authority, there is hardly a possibility for FAO and other international agencies to take unilateral action to formulate and implement programmes that help the people most in need of assistance. But the least we expect of FAO and other international agencies is to withhold support to governments that clearly have no intention to alleviate poverty and misery existing in rural areas.

'FAO should not be afraid of taking a strong – controversial – standpoint in the interests of the people of such countries. The governments that are represented in the FAO should give the FAO the moral and political support to act the way they should in the interests of these people.

'FAO will find support with us and young people at the grass-root level in the countries concerned, if they work with them instead of with the power structures that perpetuate these people's miseries.

'If FAO does not dare to be controversial it finds itself without the support of that large section of the world community which is most concerned and whose participation would be of indispensable value to the FAO.'

The recommendations of the commission were withdrawn and redrafted:

'1 Governments are urged, as a prerequisite for ensuring full participation of local people in development activities, to eliminate social injustice by ensuring equality in political power, economic wealth and social security.

2 Peoples and governments of developing countries are urged to introduce radical and full land reform to change the economic and social situation, integrated in the national development programmes and essentially directed to raising levels of living and to achieving the full dignity of the farmer. FAO and the entire UN system are called upon to mount an educational campaign to mobilize public opinion to bring pressure on governments to legislate and implement agrarian reform.

3 Governments and international agencies should recognize the important role which women have in the social and economic life of the family and the community, and involve them actively in the planning and implementation of all development programmes. To this end women's committees and organizations, as instruments for making this participation possible, should receive the support of governments and FAO.

4 Governments and international organizations should intensify their efforts to ascertain the views of local people regarding development, and should use their resources to enlighten rural people on the objectives and methods of achieving and promoting development.

5 Governments and international agencies are called upon to promote the integrated approach through coordinating technical, physical, economic, social and political factors in planning and implementation at local, regional, national and international levels.

6 In implementing agrarian reform, governments must: ensure

security of tenure; provide adequate support to develop the essential institutional framework including cooperatives for credit, supply and marketing, and advisory services; ensure that it is employment-oriented. The implementation of the total programme should be in accordance with an established timetable.

7 Governments should accept the organization of farmers and farm workers into farmers' associations and labour unions and protect the organizers and members against economic pressure and physical violence from landlords and employers. Such labour and farmers' organizations should be given the opportunity to participate in the making of all economic, political and social decisions affecting rural development.

8 FAO should establish an independent advisory and review body within the Director-General's office to study and assess all programmes and projects in terms of their possible economic, social and political consequences. FAO should strongly advocate projects which will benefit poor people even though governments may not be inclined to accept them.

9 FAO should re-orient its work and expand its resources for advice and assistance to Member Governments in implementing agrarian reform, including the strengthening of research, training and evaluation programmes and provision for more expeditious distribution of information to Member Governments.

10 FAO should promote the establishment of practically oriented education suited to the needs of rural areas, and give greater emphasis to an understanding of the dynamics of economic, social and political change.

11 FAO and other international agencies should review their internal organization and arrangements for interagency coordination in order to bring about greater cooperation in planning and implementation projects requiring an integrated approach.

12 FAO should make increased resources available to those units that deal with the social and institutional aspects of rural development.

13 FAO and other international agencies should urge governments to exert efforts to convince the elders in traditional societies that women and youth are indispensable in decision making with respect to community matters. Governments should also be encouraged to

involve youth in the formulation of national youth policies, plans and programmes.
14 FAO is requested to assist in finding support for both governmental and non-governmental programmes such as those for training rural leaders and young men and women and for creating agricultural centres and international rural youth exchanges. In providing this assistance, FAO is urged to give priority to the non-governmental agencies not enjoying government support, and particularly to the youth organizations.
15 FAO should decentralize its activities to permit more effective assistance to governments in dealing with regional and local activities. In this connection, it should provide assistance to governments in carrying out the necessary surveys and research for preparing and implementing plans for expanding rural employment and for dealing with other aspects of rural development.'

Several of these recommendations would, if adopted, alter the entire administrational policy, not only of FAO, but of the UN itself. In particular, it would encourage the participation in UN agencies of non-governmental bodies. This process has begun, but it is delicate and likely to be resisted by governments that are not over-enthusiastic about the international activities of organizations based within their own borders that are critical of them. It is unlikely that these recommendations will become FAO policy. In his closing speech to the Congress, Mr. Boerma rejected outright the idea of involving FAO in the internal politics of member nations:

'I must make it very clear that ours is not a political organization and that there are in fact very good reasons for it not being one.

'The most important of these is that it is one of the main purposes of FAO to bring together the so-called developed and developing countries to work out problems of food and agriculture. We serve, in fact, in the role of an honest broker. . . . If we were to turn into a political pressure group, we should lose that role and, with it, much of even our present effectiveness, let alone the chance to achieve a great deal more in the future.

'Having said this, I think it is only fair to add that, by the very nature of things, the success or failure of much of our work depends on political decisions. A good example of this is the need for a better and fairer distribution of incomes, which has been one of the main concerns of this Congress and is at the root of agricultural development and indeed economic development as a whole. It is only by increasing effective demand that we can hope to vanquish hunger and malnutrition. Other matters such as agrarian reform, rural employment and the liberalization of trade, which are equally aimed at economic and social equity, and on which FAO has long been at work, also require political decisions.

'The one thing that we cannot do, however, is to go around pointing the finger at individual governments. To do so would, in contemporary jargon, be almost totally counterproductive. For FAO is an intergovernmental organization. And, although I believe that it is one of the most important duties of the Director-General to take an independent stand on international agricultural policy issues in the light of what he believes to be just and necessary, he would by the same token be failing in his duties if he did so in a way that antagonized governments to the point of getting the Organization's throat cut.'

The throat is easily cut and Mr. Boerma's statement underlines the dilemma in which FAO finds itself. Perhaps there are times when an arbitrator should be critical, perhaps even of the party that retains his services. Perhaps there comes a time when it is desirable for him to climb down off the fence. Perhaps the future – not the quality of the future but the very possibility of it – for millions of human beings is such a point, and if that future is being jeopardized by the action or inaction of governments, perhaps FAO should say so, quietly at first, but then loudly and publicly. Would governments withdraw their support? Perhaps they would, in which case it might be argued that the limited resources of the Organization could be employed to greater advantage in a political climate more favourable to its views. If support were withdrawn, openly and in public, would not the government concerned exhibit itself in an unfavourable light? Would the consequent loss of face be an adequate deterrent? Has the FAO more power than it supposes, and could it blackmail recalcitrant governments into remaining within the Organization and at least continuing the discussion?

It is difficult to know and FAO's reluctance to take chances is understandable. There are, too, doubts about the ethics of trial by public opinion. On the other hand, the problems are desperate. Unfortunately, there is another and greater danger. While the withdrawal of the government of one developing country would not affect the future of FAO, the withdrawal of one of the major powers might bring down the entire UN. All governments are hypersensitive to criticism from outside, not least those of the rich countries. It is far from impossible that any official statement that could be interpreted as hostile might produce a petulant reaction. This could take the form of withdrawal of foreign aid, withdrawal of backing for FAO programmes, resignation from FAO or resignation from the UN – or a combination of or even all of these extreme measures.

It is the fear of offending any of its members that inhibits multi-organizational committees based on a concensus view of a problem. It is this same fear that confines intergovernmental agencies. When the fear of causing offence inhibits the performance of the task for which the agency was formed, then the agency would seem to be faced with two alternatives. Either it must accept compromise after compromise, try to do its best in a situation that is inherently hopeless, and in the end fail gallantly, or it must reject the inhibition, make its criticisms openly and boldly, bring specific charges against governments and against individual ministers when it is threatened and, if its very existence is made impossible then it must dissolve itself, advising the parent body and world opinion of the reasons for its dissolution and warning them of the likely consequences. Its staff might then find employment with voluntary organizations less concerned with protocol and less subject to political pressures. Either way the choice is difficult and calls into question the whole concept of the UN and the likelihood of its succeeding in resolving satisfactorily any situation that poses deep political problems.

13

Off the Horns of the Dilemma?

We are on the horns of a particularly unpleasant dilemma. The world population is rising and food production is quite unable to keep pace. The only satisfactory solution must be to limit populations, but this is the least acceptable solution so far as many of the countries most severely affected are concerned. The present aim, therefore, is to produce more food. All the schemes advanced to this end so far have failed and are failing – including the 'Green Revolution.' Whether they fail this year or next, in the end it is clear that fail they must, for endless expansion is impossible within a finite system. The main difficulty at present lies in persuading most of the peoples of the world – and particularly those living in the developed countries – that the system is finite. We have been brought up in the shadow of the nineteenth century belief in the omnipotence of man and the limitlessness of the resources available to him.

Is disaster inevitable? Have we exhausted all the options before us so that now there is nothing to do but sit back and wait for the end? Bad though the situation is, it is not quite so bad as that. There is hope, provided we accept that without a reduction in world population no long term solution will work and provided we accept also the immediate need for radical reform in many developing countries together with a reduction in the material standard of living in the developed ones. Wealth and resources must be distributed more evenly throughout the world. This is a matter of basic justice. It is also a matter of the survival of all of us.

At the present time the economic inequalities between rich and poor are widening and with the world's present economic systems it is inevitable that they will continue to widen. There is a business game that children play which demonstrates very clearly that if one player begins with more assets than the others it is not difficult for him to maintain his advantage and that once the disparity reaches a certain point, nothing his opponents can do will prevent him from growing richer while they grow poorer. The system itself is taking over to produce this result.

There is only one way to break the vicious circle and redress the balance quickly. The rich must give – not lend, *give* – money to the poor to enable them to develop their food production, their industries, to implement social, political and economic reforms and education programmes within their own territories and to institute population programmes and other measures that will lead to stability. This is not a matter of collecting tins on the bars of our public houses. What is needed is a series of major contributions from all of the developed countries. The FAO has suggested a modest 1 per cent of the Gross National Product (before that it was 1 per cent of the national income) from all the industrial nations. The suggestion was made several years ago and it was ignored. As is the way with ignored bills, the price has risen and now it is doubtful whether 1 per cent would be anything like sufficient. This is not a request for charity, it is not a donation made by a generous squire to a deserving cause, it is a debt or, if you prefer to look at it another way, an investment in the future of humanity.

Anything less than a major redistribution of wealth tends to benefit the rich more than the poor. Aid given to finance an industrial development project, for example, will pay for a company in one of the rich countries (usually the donor country) to build factories in the recipient country. Both governments and the company itself will have the very best of intentions. The aim is to benefit the poor. The factories create employment, so raising standards of living locally. Local personnel are trained to take over the running of the factory at an agreed date in the future, but by the nature of the system, the profits from the factory must be used to pay dividends to shareholders in the donor country. Thus the enterprise takes profit, which otherwise might become capital, from the recipient country and returns it to the donor country. Even food aid, sent to relieve an emergency, usually represents an embarrassing surplus that cannot be released on to the world market for fear of creating a glut.

As it is, much of the finance that reaches the FAO is spent on aid to relieve disaster areas. It is spent on food, bought from the rich to give to the poor and hungry. It is administered by the FAO's World

Food Programme. Of a total of a little over 725 million US dollars that has been donated to the WFP since its inception in 1963, and including pledges for 1971–2, 520 million dollars has been spent on commodities and 205 million on cash grants and services. A further 47 million dollars has been donated in the form of food grains. In April 1971, for example, 35 million dollars was spent on wheat to establish a wheat reserve in Turkey. During the spring and summer of 1970, 2,500 tons of wheat was supplied to the Paktia province of Afghanistan to make up for deficiencies caused by a poor harvest. An emergency grant of 25,000 dollars was made early in 1971 to provide wheat flour, dried skim milk and edible oil to the victims of floods in rural areas of Bolivia. 492,000 dollars was spent in the same period on sorghum and maize for the victims of a drought in Cameroon. A daily ration of millet and vegetable oil sufficient to last 20 days was airlifted into the interior of Chad to feed people displaced from their homes by civil unrest. Drought in Kenya brought 800,000 dollars worth of dried skim milk and vegetable oil – to feed 140,000 people for six months. Much of Africa was affected by drought in 1970. The farmers of Mali lost most of their sorghum, millet, maize and rice crops, creating a deficit of 80,000 tons in the staple foods of the country. The WFP contributed 492 tons of sorghum, 979 tons of maize and 194 tons of wheat flour to be distributed over 120 days among 45,000 people, with supplies for pre-school children and nursing mothers for 250 days. The cost to WFP was 243,000 dollars.

These are typical examples of WFP administered aid to disaster areas and they all took place in April and May 1971, at a total cost of 1,560,000 dollars. It is essential, of course, to provide emergency supplies in this way, but it is a stable door operation, rescuing people after disaster has struck.

Nevertheless, there is a more positive side to the work of WFP, a side that gives some reason for optimism. Frans Deeloman, Director of Agricultural Assistance to Developing Countries in the Dutch Ministry of Agriculture and Fisheries, summed this up during a speech he made early in April 1971, to a session of the Intergovernmental Committee of the WFP. After mentioning the other UN agencies concerned with development, he went on to say:

'But WFP is a unique programme, a food aid programme which

should be geared to the unique needs of the people who need food. Money of course is the best kind of development aid, but with food it is possible to devise projects which aim directly at those groups which need food. And we must keep in mind that those who need food are those who are politically less powerful in the developing countries. WFP is therefore an excellent instrument for the promotion of *additional* social and economic development. As a multilateral organization it is well filled to try to reach those classes of the population which bilateral programmes are often not able to reach.'

The use to which food aid can be put is illustrated by a project the WFP undertook in Kizilcahamam, in northern Turkey. The farmers were in a desperate plight. The forests had been stripped from the hills and the rains were washing away the top soil into the river, which carried it as far as a dam below the village, which was silting up. The erosion was aided by the 85,000 goats that destroyed what vegetation they could find. The 17,000 cattle were poorly fed, their milk yields were low and their calves fetched low prices in the market. Early in 1964, WFP, at the invitation of the Turkish government, began what it called 'Project 099.' Rations were brought in for the farmers and their families to prevent them from leaving the area and as an incentive to them to join in the work of restoring the forests, planting trees on the denuded slopes and improving the pasture. Sheep and more cows were brought in in exchange for the goats. Slopes were terraced and eventually a new, better-yielding breed of cattle was introduced. Today there are twenty-six villages and 40,000 people involved in the project. Feed for the cattle was brought in to protect the pasture while it became established. Now there are less than 6,000 goats and their grazing is strictly controlled. Erosion is ended. In the valley they are growing alfalfa (lucerne) to feed 1,000 of the new cattle, many of which are into their second generation. The net income of the farmers has increased by up to eight times and the villagers are no longer so heavily in debt to moneylenders. The whole project has cost 544,000 dollars.

On 30 March 1971, Francisco Aquino, Executive Director of the WFP, proposed to the WFP's governing body that it should seek donations of 400 million dollars in food, cash and services for the years 1973 and 1974. 200 million dollars of this would be spent on development programmes – like the one in Turkey. The target was supported strongly by Mr. Boerma and by Mr. W. van der Heide, who attended the meeting on behalf of the UN Secretary-General. At that time – the end of March – WFP resources for 1971 and 1972 amounted to 192.8 million dollars, a figure they had just reached with a pledge of 9 million dollars from Sweden.

Money is hard to come by. Britain has pledged 6.276 million dollars for 1971 and 1972 (less than Sweden), West Germany 1.5 million. Also on 31 March, the WFP appealed for 2 million dollars to feed 1.5 million Palestinian refugees. This followed a request from another UN agency – UNRWA – which faced a deficit of 3.7 million dollars. A few days earlier it had been announced that WFP would send food aid for three years to 100,000 people – one-seventh of the population – in Guyana, bringing the total WFP aid to that country to 764,000 dollars over three years. Altogther there are now 10 million people in sixty-three countries that receive food aid from WFP.

Of course it is essential that food be held for distribution in the event of disaster. This is no more than sensible planning on the part of governments and the international community. It is no answer to the world food problem. Indeed, it is irrelevant to it, and it is dangerously misleading to suggest to individuals who donate small sums to charitable organizations or to the citizens of those countries whose governments contribute to WFP, that such donations are in any way helping to satisfy the hunger and malnutrition of those who are chronically under-fed.

Mr. Boerma made this point in a speech to the 25th anniversary meeting of the International Federation of Agricultural Producers in Paris in May 1971. He appealed to governments to make use of the existing international machinery to correct what he called the 'completely unbalanced' state of world agriculture. He was referring specifically to the fact that at the present time farmers in the developed countries are bedevilled by surpluses, while those in developing countries cannot produce enough. This is due to more favourable climatic conditions in most of the developed countries, coupled with sustained higher investment in a more prosperous economic climate and access to the best research facilities, all of which are applied to farming in the developed countries. Yet it works against the poorer farmers by keeping world food prices at a level that is too low for them to be able to produce and export profitably. Mr. Boerma pointed out that food and agricultural policies are now a matter of global concern.

'The level of national agricultural production in any one trading country automatically affects the situation in others,' he said, 'Competitiveness in agricultural exports is more and more coming to mean, in the case of many products, the ability to compete in the provision of export aids, or, for those countries who cannot afford to do this, the ability to accept low farm prices and farm incomes in order to maintain their export supplies on world markets.'

One of the requirements of agricultural development is to raise farm incomes and levels of rural employment so as to create prosperous communities. Without this, stagnating economies must continue to stagnate. Without higher wages the hungry cannot afford to buy food, no matter how much of it there is. Unless there is a market for the food, farmers cannot produce it. Just as any improvement within countries must depend on a general level of prosperity that will provide a market, so in the international community, countries must be able to increase their national prosperity. This is not a plea for endless growth on the pattern of the developed countries. It is a recognition of the need for developing countries to advance beyond their present position. The rich must be prepared to forego some of their profits, which in any case are excessive. There is no lack of machinery for achieving this. What is lacking is the goodwill of the governments of the rich countries. Mr. Boerma deplored this lack of political will in view of the grave world agricultural situation. He said:

'I therefore appeal to them with all the force at my command to take urgent steps to resolve the differences that exist among them on their national agricultural production and trade policies, and thereby to salvage the interests not only of their agricultural communities . . . but also their own interests and those of all the people that they govern.'

If the rich countries can be induced to part with some (not all – no one is asking for that) of their wealth, then the developing countries may have a real chance to feed themselves.

How should this aid be spent? This is a matter of constant debate. There can be no doubt that for a number of years the last people to benefit from aid were the poor and hungry. Aid was a pawn of international power politics, a weapon of the cold war. It was spent on magnificent buildings and prestige projects: we have considered the Aswan Dam already as an example of such a project. It was delivered in the form of large machines, brightly painted, with 'Gift of the United States' or 'Gift of the Soviet Union' emblazoned on them. They were big and gaudy and impressive and they showed the world, particularly the people of the countries they were meant to impress, the power and magnanimity of their donors. They were totally irrelevant to the needs of the people. In most cases they rusted away in obscure corners for want of spare parts, trained mechanics to service them or even skilled operatives.

This was a situation that Dr. E. F. Schumacher, an economist with the British National Coal Board, set out to rectify. He has travelled widely in developing countries and he has been an economic adviser to several. He saw the tractors and the combine harvesters that could not be used. He saw, too, people growing less food than they might for want of a better plough, going thirsty for want of a better well. He founded the Intermediate Technology Development Group Ltd. (ITDG). The ITDG concept is based on a recognition of the need to increase employment in rural areas by the introduction of technologies that will provide people with the capability to increase their productivity, and so their prosperity. The technology, in fact, must be appropriate to the level of economic development of the recipient community. In the words of ITDG itself:

'We are talking about technologies which raise the productivity of labour by methods that make the maximum use of the skills and initiatives of the people and require the minimum of capital – that is, technologies which make possible the mobilization of labour power (unemployed and underemployed rural dwellers in developing countries) into productive activity. This implies that such techniques must be relatively cheap, relatively simple and labour-intensive, and make maximum use of indigenous skills and materials. They are "intermediate" because they will be more sophisticated and expensive than those which the poor are using now, but less costly and complicated than the machinery employed in the highly industrialized countries. Dr. E. F. Schumacher put this in his original paper in terms of "average capital cost per workplace." This is a useful approach because it enables one to see that the concept of intermediate technology is dynamic; as people become richer and more skilled they can afford, and make and utilize, more complex and expensive equipment.'

Dr. Schumacher has said that to be really effective, tools and machines for rural areas in developing countries must be able to pack into the back of a Land Rover, so that they can be taken anywhere, and then they must be so designed that the villagers can assemble, operate, maintain, repair and eventually replace them. The tools must be tough and versatile. ITDG promotes, for example, a tool bar that can be towed behind a tractor, or a horse, or oxen. There are several different tools than can be attached to it and that form part of the kit – two kinds of plough, a rotavator, a harrow, discs. The whole outfit is made of steel. It is simplicity itself and as parts wear out a competent village blacksmith can repair them. The cost of the complete kit is around £60. It is easy to see the relevance of such equipment compared with that of a machine that costs £5,000 and which requires one operator, who is paid probably £120 a year, and which may well put three more men out of work.

ITDG is not universally popular, for there are still governments in both developed and developing countries that are devoted to the large, prestigious projects and there is still a belief that it is possible to introduce advanced technologies in areas and among people who are not and cannot be trained to comprehend the concepts behind them.

ITDG assembles information and makes it available to whoever needs it. It finds out where the cheap, simple tools are being made. It discovers what is needed and finds a factory that will manufacture it. To support this activity it sponsors field projects to test the validity of its concept and to experiment with intermediate technology and ways in which it may be introduced.

Now, if there can be a genuine redistribution of wealth, and the capital so liberated can be applied in the ways that are most appropriate to the needs of the people in the developing countries, progress really can begin.

There is one more need, however. No matter what help may come from the developed countries there must be co-operation within the developing countries. Some authorities believe that reform of the internal political and economic structures of recipient countries is the most urgent task. It is also one of the most difficult, for rich countries are curiously sensitive about manipulating their aid in order to modify the internal policies of other governments. This is curious because while they are so shy of attaching strings to aid that would benefit the populations of the recipient nations, they have not the slightest hesitation in reacting sharply and often brutally if those nations attempt to assert their international independence and especially if they carry this independence of thought to the length of taking control of resources within their borders that the rich countries regard as properly theirs. The relationship between the United States and Cuba, once Cuba attempted to overthrow American economic domination is a case in point, as is the reaction of the United States to the attempt by the people of Vietnam to appoint a government that was popular with them, but unpopular in Washington, and which might have removed from the American sphere of influence important natural resources. In the end, though, the attachment of strings to aid may be the only quick way of bringing about vital reforms. This is the view of Gunnar Myrdal, the Swedish economist. He believes aid should be used directly as a force for social and political reform as well as economic growth. His vision, he has said* is to 'get the rich countries to overcome their present polite reluctance to touch on the need for reforms in the underdeveloped countries' and, 'the rich countries should find a common cause with the liberal forces in the underdeveloped countries who press for reform.' Professor Myrdal believes that the whole UN development programme is doomed to failure unless these reforms are carried through.

What reforms? We saw in the last chapter that land reform, especially in Latin America, could increase productivity from farm lands. Each year the FAO makes an award to the field officer who, in the view of FAO, has contributed most to the country to which he or she is assigned. In 1970, one recipient of this award was Dr. Solon Barraclough, an American who has worked in Latin America. He has defined land reform as 'giving the landless and nearly landless rights to good farm lands now held in large estates.' He went on to

* In *Ceres,* March 1971, published by FAO, Rome.

say, 'At the heart of all Latin America's rural problems is the land tenure system. Too few own too much and they usually have all the power as well. Few people in rich industrialized nations realize how important control of the land is in agrarian societies. If you control the land, you control everything else.

'In Latin America land ownership is tied to the power to make others do one's will. So the first step, though by no means the last, in solving the problem is redistribution of the land itself.'

'In Latin America half the inhabitants are, still very small farmers, tenants or landless farm workers. They are poor by anyone's standards. Despite the rapid spread of modern technology, there are more peasants and landless workers now than there were a decade ago.'

Dr. Barraclough is critical of plans of the 'Green Revolution' kind that aim simply to increase yields through improved technology. If they succeed, such schemes can actually make things worse for the very poor. 'There is a dangerous misconception in the world today,' he has said, 'Development is not merely economic growth. It is not just increased output. Real development means an improvement in the quality of life. Income has to be more equitably distributed. Justice must be accessible to all, on equal terms for rich and poor alike. People everywhere must feel, indeed *know*, that they are free to participate in determining their own destiny.'

The FAO is well aware of the need for agrarian reform and it has a committee devoted exclusively to this problem, appointed in 1970 on the recommendation of the UN Economic and Social Council, the ILO and FAO itself. Currently it is helping fifteen countries with research and consultants to governments. We should not underestimate the difficulties involved in a successful scheme for land reform. It is not enough to introduce reforming legislation if this simply leads to the old landlords being replaced by new entrepreneurs – and this sometimes happens. Dr. Barraclough, who, apart from his work for FAO, is professor of agricultural economics at the University of Cornell, points out that whole social structures may have to change.

'Changing the social structure, however, is a serious and controversial matter. Existing power relationships are altered. The distribution of wealth and incentives are affected. Even such an innocent-appearing new institution as an agricultural bank or modern marketing system may pose serious threats to the powerful established interests. Land tenure changes are resisted everywhere by the traditional élites and by many other groups.'

The reform of land tenure may be accompanied by other reforms that may release the stranglehold of the moneylenders that are so often among the principal beneficiaries of aid programmes, and it may reduce the amount of corruption that erodes any form of aid in many areas by diverting individually small, but collectively large, sums of money into the pockets of merchants and bureaucrats all the way from the port to the village.

We have considered, too, the increased food availability that would result from a reduction in the food that is wasted after harvesting. There are other forms of waste that could be reduced, also.

The most glaring of these is the waste of plant nutrients. The pollution of water from agricultural fertilizers and from sewage is a serious problem in the developed countries. Development on the pattern of those countries will result in an inevitable increase in water pollution which may be even higher in the case of tropical countries where soils are poorer and where the expanding urban areas have even fewer facilities for treating sewage, and the result may be more serious still, owing to the more extreme pressure on available fresh water resources in many of those areas.

A great deal of the work that has been done in Israel is of immediate interest to other developing countries. Systems of irrigation have been introduced there that are based on piping water, which may also contain fertilizer, underground to the root area of the crop plants. This reduces surface evaporation and the consequent accumulation of salts in the topsoil. Now there is news of another project which employs denitrifying bacteria located deep in the topsoil of certain areas. These are used to break down nitrogen compounds, liberating the gaseous nitrogen to return to the atmosphere. The technique was developed to correct the eutrophication of the Sea of Galilee, which is now called Lake Kinneret and which is Israel's main reservoir. Here the problem is caused by the draining and reclamation for agriculture of a former inland lake and salt marsh in the Hula region. The soil that is exposed to the air for the first time has an abnormally high organic matter content – somewhat similar to certain peat fenland soils in England – and the soil is oxidizing to carbon dioxide, water and nitrates. The air penetrates only 30 to 40cm below the surface and beneath the oxidizing layer there is a further layer which is in an anaerobic, or oxygen-depleted, state. It is there that the anaerobic, denitrifying bacteria are found.

This is an unusual situation, but it offers possible hope for the treatment and recovery of water suffering from eutrophication. The treatment itself is perhaps more interesting. The affected water is pumped back to the land and sprayed on the surface. As it penetrates to the lower levels it loses its nitrates. Does this suggest that run-off water might be returned directly to the land to be recycled not only for irrigation purposes but also to achieve a more efficient use of the nutrients it contains? All water that runs off farm land contains a proportion of plant nutrients, whether the land has been treated with artificial fertilizers or not. Recycling might offer a sensible economy of the nutrients that are available.

Except in areas where there are intensive livestock units, pollution from untreated sewage is an urban problem, and one which every civilization has experienced. It, too, can be solved. It needs to be solved, for, as we have seen, the pollution is only half of the problem. The other half relates to the plant nutrients and organic matter that is removed from the land by cropping and that is not returned. This represents a breaking of a biological cycle. In the rich countries the organic fraction of urban wastes, excluding sewage, is increasing. According to Dr. K. R. Gray, K. Sherman and A. J. Biddlestone* it will reach 70 per cent in Britain by 1975.

Organic material of this kind, and particularly sewage, requires treatment before it is used on the land. Where sewage contains industrial effluent, there may be levels of metallic residues that are unacceptably high. This is a problem that can be solved, however. There is a disease risk from pathogenic organisms in the material and also, applied raw, decomposing organic matter can actually reduce the amount of organic matter present in the soil. This is because the fermentation process by which it is broken down by aerobic bacteria into simpler chemical compounds can affect other organic matter if it is too vigorous. The organic matter in the soil is oxidized – or 'burned off.' Obviously, the material is unpleasant to handle in its raw state and this is a third reason for treating it. Not

* A review of composting, *Process Biochemistry,* June 1971.

only is it offensive, it is also bulky, and that makes it expensive to handle and to transport.

However, this urban organic waste matter can be composted on an industrial scale to produce humus, a light friable, fairly dry and pleasant substance with no smell and containing no pathogens. Essentially, the process accelerates what happens in nature. It resembles a sophisticated garden compost heap. The material is prepared by sorting out inorganic and non-compostable materials and by pulverizing the compostable portion. This is placed in large metal drums and partly dried sewage is added. The drums rotate slowly so as to keep the pile moving and the temperature is permitted to increase rapidly to about 60°C. As the temperature rises, and again as it falls, different groups of bacteria take over the fermentation process until, after a few days, or at the most weeks, the organic material is converted into humus. Weed seeds and harmful organisms have been destroyed by the sustained high temperature and by antibiotics produced by fungi. The pile is then removed from the drums and left to mature at ambient temperatures for several months, during which time decomposition continues until the final product is a valuable soil conditioner containing a wide range of beneficial soil micro-organisms that will facilitate the release of plant nutrients. The composting of the wastes of, say, Calcutta, might supply an invaluable soil stabilizer and source of fertility for the farmers attempting to feed that city.

Unstable soils can be stabilized. There are a number of processes that achieve this – one uses latex sprayed on to the surface. The problem on an unstable soil may be to establish plant cover in the absence of humus. Willy Krohnstad, a Swedish engineer, has invented, and is now marketing, a product he calls 'Essbinder,' which is an organic material, manufactured from the wastes of the Swedish paper mills. It is packed in 200 litre drums, which are easily transportable, and before use it is dispersed in water to give a solid matter concentration of 65 per cent. This is sprayed on to the surface where it forms a thin layer which congeals and prevents evaporation of water, while permitting rain to percolate through. The film is porous. This is important, because in some areas the major source of fresh water is dew, which can thus be retained. At the same time this protective layer conserves heat in the surface layers and also prevents wind erosion. By preventing the evaporation of water from the surface it prevents the build-up of inorganic salts that might be caused by irrigation, so making irrigation easier and safer. Both by retaining water and by conserving heat it increases seed germination rates.

Of course, it is expensive, as is the municipal composting of refuse, but we have suggested already that if the world food problem is to be solved, then there must be aid from the rich countries on an unprecedented scale, and a major diversion of global resources to this end.

The aim of recycling organic wastes and stabilizing unstable soils is to increase the organic matter content of farm lands. Over a period of years this will enhance their fertility and increase the efficiency with which artificial fertilizers can be used. Therefore investment in these areas will be likely to result in a healthy, stable and productive agriculture. The soils will be enriched, whereas a rapid conversion to high-technology farming, even if this were possible, would impoverish them in the long term. One approach aims at increasing biological stability, the other undermines it.

What we are saying, then, is that the problem can be solved provided the world's wealth is redistributed, provided that the increased investment in developing countries is used, partly, to reform radically those political, economic and social systems which inhibit the efficient use of available resources and that the world's reserves of materials be redirected. Philosophically, we need to accept the fact that no matter what life looks like in London or Washington, or Moscow or Tokyo, we live in a predominantly agrarian world. Most of the world's population is composed of country people. Urbanization has gone far enough. It is time to reverse progress, to regard people as more important than buildings and farms as more important than factories. In the final analysis, you cannot eat machine tools.

Can it be done? Can it be done without world revolution? How else are we to persuade the governments of the developed countries that they must, to use Prof. Ehrlich's word, 'de-develop?' How will we persuade Parliament that Britain is grossly-over-populated already? There is some slight hope that a growing grass-roots movement, especially among young people, rejects materialism and realises that the survival of man depends on the rapid evolution of changed attitudes in the developed world. Will they be able to inform and educate a sufficiently large proportion of their own populations in time? Will they convince their governments, which are committed to economic growth and which refuse to admit that a population/resources problem exists? Or will they replace those governments?

What, then, of the developing countries themselves? There, in many cases, governments and forms of government must be replaced. Existing social structures, disrupted already by over-exposure to western values and ideas, must be overturned. Can it be done? It seems unlikely that it will be done in time without the active co-operation of the developed countries.

If it can be done, what kind of a world is it that we are creating? If steps are taken to bring populations into balance with resources, and if drastic action is taken to improve the immediate situation in order to buy time, there is no reason why a smaller, stable population cannot be fed indefinitely.

There is one country that has attempted to deal with the problem. China, one of the largest countries in the world, and a poor one, has accepted that its major resource is its people and that its society is, and should remain, agrarian. Farming is labour-intensive and every citizen is required to contribute labour, directly or indirectly, to the production of food. Nothing is wasted. Everything that can be returned to the soil is returned. Human excrement is bought and sold! Earth is moved in baskets. The production of artificial fertilizers is small and they are not generally popular, because the over-use of them is known to be bad for the soil. Pesticides are used, but on only a small scale, and then only the less toxic ones, for a major agricultural occupation is the rearing of fish in lakes and ponds and pesticides such as DDT – which is not used – are known to kill fish. All industrial effluent is recycled, so that pollution of the environment presents no serious problems. Power is provided by small stations operated in series by the movements of river water. People are poor, the country is poor, but, for the first time in its long history, China is no longer torn asunder by famine and civil war. With 700 million people, the Chinese do not consider they have a serious population problem and they may be right, for efforts to stabilize it appear to be successful and the growth rate has been reduced. Should they need to reduce it they may be able to do so.

This is not to say, of course, that the social organization and the way of life that prevails in China is exportable, certainly not beyond those adjacent countries that have been under Chinese cultural influence for most of their history. It does show, though, that without a high level of technology, without many machines or agrochemicals, food can be produced and distributed to give an adequate diet to a large population. However, in order to reach this position, China had to undergo the most radical and, in terms of human survival, the most significant revolution and subsequent political reorganization this century has seen. It is no coincidence that it is to China that many of today's young people look for inspiration and hope.

Just as it is possible to make projections of what life might be like if populations are allowed to continue to rise, if resources are mined out, if the polarity between rich and poor intensifies, so it is possible to guess at what an alternative civilization might be like. The future is not entirely outside our control.

Imagine, then, a world in which populations were stabilized at a level lower than today's. Imagine that the planet's wealth and resources were distributed evenly. Imagine that the production of food were geared to a labour-intensive, low-input high-output system of husbandry. What might life be like in Britain?

Most of the population would work on the land for at least part of their time. Cities would be much smaller than they are today. Government would be decentralized, so that decisions affecting the day-to-day life of the people would be taken by regional councils in which the people themselves would participate. Food would be plentiful and of high quality, although it might include items unfamiliar to us today and diets would be likely to contain less meat. Power would be generated on a small scale by the harnessing of solar and gravitational energy. There would be comparatively little industry and what there was would be largely automated. Work would be pleasant – there would be no drudgery. Indeed, work might be voluntary and optional. Durable goods would be made to last for long periods and when eventually they wore out the materials from which they were made would be recycled. There would be little private transport, but improved public transport systems would provide such mobility of people and goods as was required. However, people would tend to stay in one place, since there would be no need for labour to be moved about the country. For this reason, family groupings would be larger. The present nuclear family, of parents and children, would have been replaced by one or another form of extended family. Instead of being spent on material possessions, wealth would be spent on goods, services and activities that consumed the minimum of resources and had the minimum impact on the environment. Education, literature and the arts would flourish. It is very probable that there would be a new major religion and that a main tenet of this religion would be respect for other living things and life processes. Ritual and ceremonial connected with the religion might occupy a good deal of people's time. There would be parks and nature reserves, theatres and concert halls, social centres and sports facilities.

Does this sound Utopian? All of it is possible. Will it happen? That is another matter.

14

Who will eat?

What are we to make of it all? In the end, who, if anyone, will eat? Are we headed for catastrophe or paradise, will the rich continue to grow richer while the poor grow poorer or will the world's riches be shared more evenly across its surface? Will we create a brave new world out of the hunger the misery, the war and the social chaos? Or will the world remain much as it is? Or will it collapse within the lifetime of most of us?

The arguments sway back and forth, each side calling up facts – indisputable in themselves – to support one or other extreme position. In his presidential address to the 1971 conference of the British Association for the Advancement of Science, John Maddox, editor of the scientific journal *Nature* and a noted optimist, denounced what he called the 'doomsday syndrome.' Too much attention to the environmental problems which may or may not be facing us could actually deter scientists from dealing with them, he suggested. He believed the world can continue to feed its growing population. His view is supported by A. M. Edwards and Prof. G. P. Wibberley, who maintain* that Britain is not overcrowded and will not be overcrowded with the 60 to 70 million population that is forecast for the end of the century. There is ample land, they say, and provided that its use is planned properly, and agriculture can adapt to feed the additional numbers.

At the present time, however, the optimists are heavily outnumbered. Edwards and Wibberley appear to believe that provided there is land for urban development and recreation, no other problem exists – an extraordinary view to emanate from one of the country's leading agricultural colleges. Agriculture is becoming more and more productive, they seem to argue, so that by the end of the

* *An Agricultural Land Budget for Britain,* School of Rural Economics, Wye College, 1971.

Gross Agricultural output per hectare of agricultural land and per active male in Agriculture

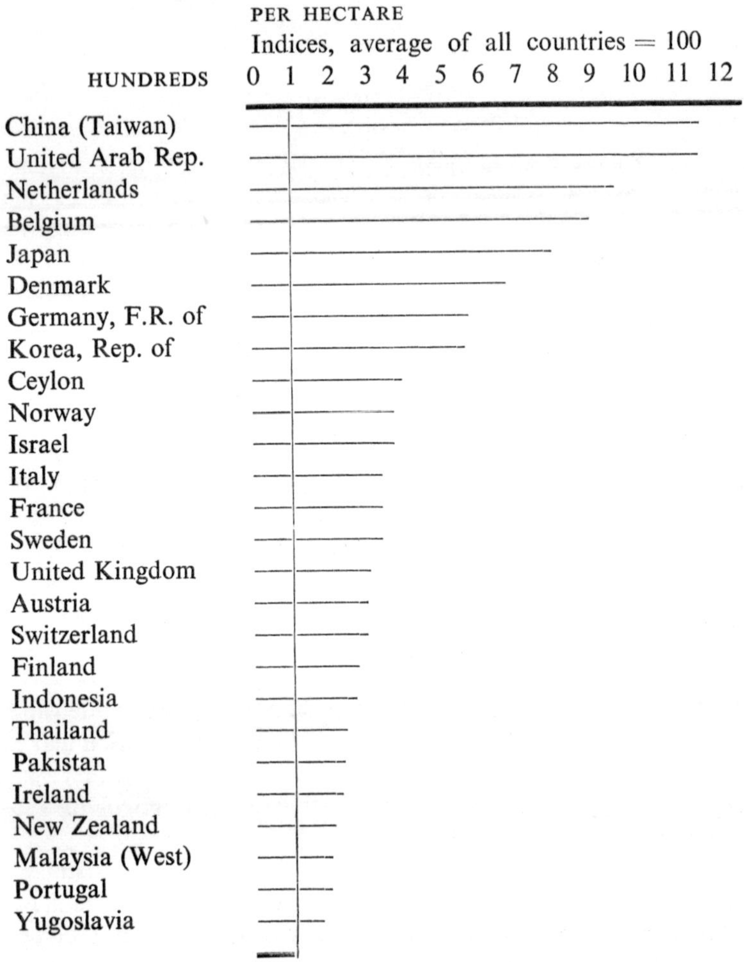

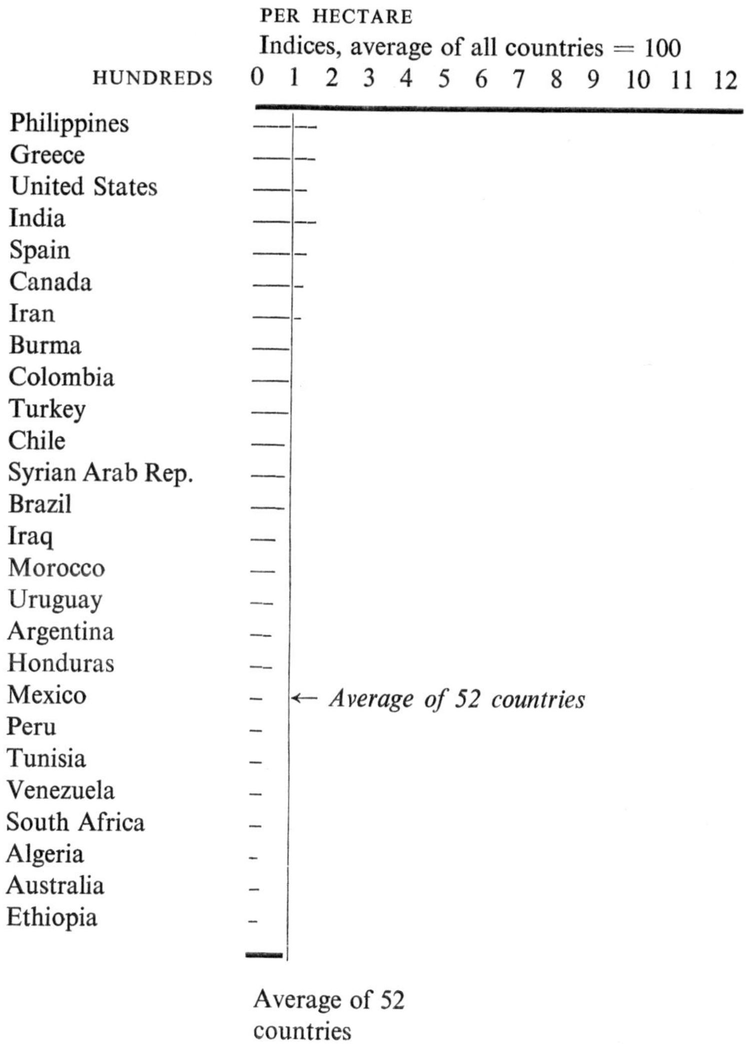
PER HECTARE
Indices, average of all countries = 100
HUNDREDS
0 1 2 3 4 5 6 7 8 9 10 11 12
Philippines
Greece
United States
India
Spain
Canada
Iran
Burma
Colombia
Turkey
Chile
Syrian Arab Rep.
Brazil
Iraq
Morocco
Uruguay
Argentina
Honduras
Mexico
← Average of 52 countries
Peru
Tunisia
Venezuela
South Africa
Algeria
Australia
Ethiopia
Average of 52
countries

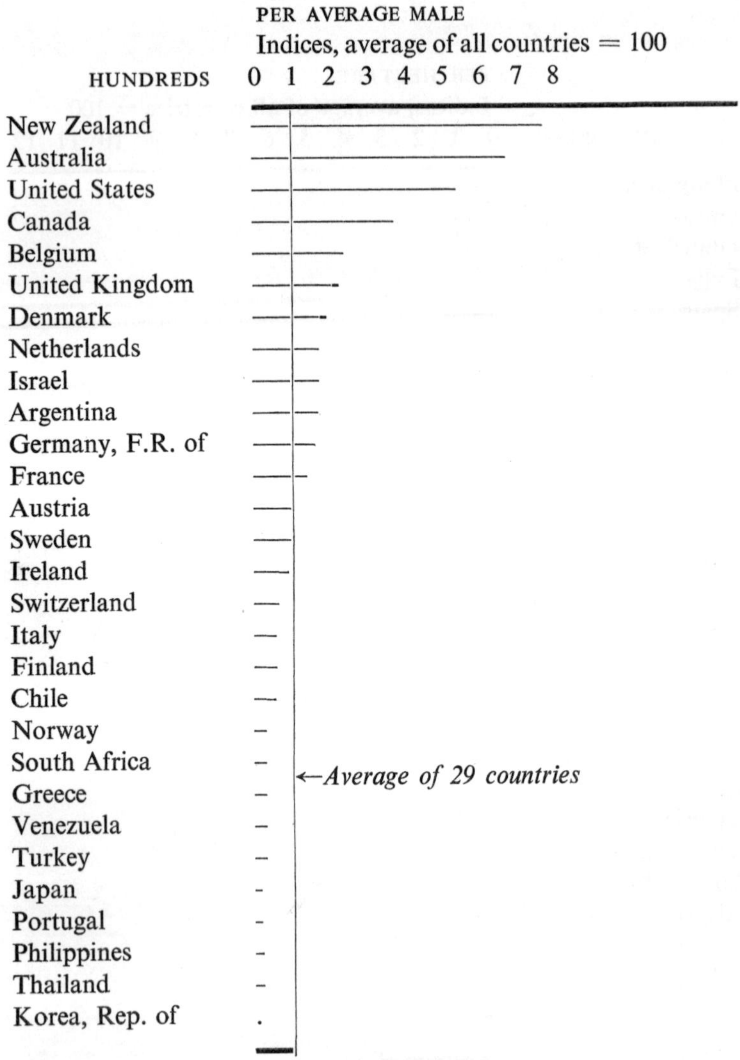

Gross agricultural output per hectare of agricultural land compared with output per active male engaged in agriculture. Data for gross output are based on the FAO

century it will be able to feed 60 million people on less land than it uses at present to feed 30 million.

Mr. Maddox, too, came under heavy attack. Dr. Hilary Rose, of the London School of Economics, was far less confident of the future. She suggested that one day Mr. Maddox might find himself announcing a new magazine, *The Pangloss Journal*, for the study of public reassurance. Indeed, when one examines Mr. Maddox's argument in detail it is remarkable for what it omits or glosses over and it is possible to see in it the almost religious faith in the '*deus ex machina*' that is so characteristic of the more extreme optimists.

Mr. Maddox believes that the development of new farm land could feed a further 15,000 million people without any further agricultural advance. What he overlooks is that the farm land does not exist. As we have seen, there is little scope for further expansion into marginal land and this is an approach the FAO has rejected. He believes that the agricultural techniques practised in North America, if transferred to the developing countries, could feed 10,000 million people by the end of the century (where does he derive his numbers from?). Apart from being agriculturally impossible, and apart from the fact that the world lacks the mineral and particularly the petroleum resources to permit it, and apart from the fact that it would aggravate problems of rural unemployment and so be self-defeating, this belief overlooks the fact that in terms of productivity per acre the most efficient agricultural country at the present time is Taiwan, followed by the United Arab Republic, with the United States 29th in the table and Canada 32nd, behind Korea, Ceylon, Indonesia, Thailand and Pakistan. India has a higher productivity per acre than Canada – and still it is not enough. It is difficult to see how the situation could be improved by substituting a less productive technology.

---

index numbers for agricultural production. These exclude feed, seed and waste and are aggregated by using average regional producer prices relative to wheat. Data of output and agricultural area refer to the 1962–6 average. Data for males employed in agriculture are derived from censuses for 1960, 1961 or 1962. The chart shows that mechanisation and the introduction of labour-saving chemical techniques improve the efficiency with which labour is utilized, but do not necessarily increase overall production. The inference is that men are more efficient than machines and chemicals.

Nevertheless, Mr. Maddox does have a point. There is a danger that if the problems are publicized too widely and if too gloomy a picture is painted of the future, people may react by ignoring it, rather in the way that they ignore the violence of war when it is shown on their television screens day after day. Eventually they cease to see it, or it becomes meaningless. Yet what if the future really is gloomy? Should we turn away and hope things will come out right in the end? There is no reason to suppose they will, but there is reason to hope that crisis may be averted, even yet, if enough people are aware of it.

The need for population control is becoming widely recognized, even in Britain. Sir James Taylor, chairman of the Royal Society of Arts, told a joint meeting of the RSA and the Research and Development Society in London on 28 September 1971, that 'I think it doubtful whether environmental problems can be solved without population reduction or, at least, by reduction of the rate of increase ... Further, it is almost certain that the conservation problems will remain obdurate unless the individual demands on world resources are diminished.' In the United States, according to a report from the Washington Centre for Metropolitan Studies, the birth rate may be declining and may reach zero. It found, in 1971, that there were 15.5 per cent fewer children under five years of age than there had been in 1970. If it is true, this would be the first decrease since the depression and the largest decrease since 1850, when records were first kept.

Yet it is extraordinary that we should find ourselves having to prove something so obvious as that there is an upward limit to the number of people the planet can feed. Has our faith in technology made us completely blind to the finite nature of a sphere orbiting in hostile space? Once we accept that a limit must exist, then all schemes that aim to increase food production indefinitely can be seen, not only as impossible demands for a kind of biological perpetual motion, but as irrelevant. Each of them can be challenged simply by asking, 'What then?' When you have fed the 200,000 millionth human being, do you draw the line and say enough, no more? If then, why not now? No, what is really relevant – all that is really relevant – is whether we can feed the population the world has at the present time and whether we will be able to continue to feed it in the years immediately ahead as it continues to grow. The FAO recognizes this. It plans no further ahead than 1985. As regards the rather sterile arguments between the 'optimists' and the 'pessimists,' unfortunately, the finite nature of the system within which we live must defeat any attempt at permanent growth. The 'pessimists' need no more than this one fact and their case is won.

A growing population must be fed from less land, as farm land is taken for urban development. If population growth is accompanied by increasing industrialization, even more land is lost. There is concern in the FAO that the developing countries should not follow the pattern of the industrial nations and allow their cities to sprawl into fertile land. The subject is to be raised at the UN Conference on the Human Environment to be held in Stockholm in 1972. The UN Intergovernmental Working Group on Soils met in Rome in June 1971 to begin preparations for the Stockholm Conference, and listed road building, airport construction, pipelines, artificial inland waterways and other requirements of an urban industrial society as predators of agricultural land. Dr. Rudy Dudal, Chief of the FAO's Soil Resources Development and Conservation Service, told the meeting that 'Immeasurable human effort and resources have gone into maintaining this fertility over many centuries. Now this valuable land is being lost very rapidly, and without much thought of public discussion, because planners, economists and administrators tend to see land mainly in economic terms. It is a product of an increasingly urban mentality.' Miss Katherine Kenyon, in a paper *Why Does the Countryside Matter?*, published by the Council for the Protection of Rural England, sums up the effect of this mentality as it is experienced in Britain:

'We talk of landscaping and beautifying the country; our forbears talked of improving it.'

'In modern parlance trees are called screens, and in evidence given at some public inquiries appear really to be regarded as static furniture. The word preservation is often shied at; technology and progress (to what? more wealth?) have made it suspect, comparable to the aims of a museum . . .'

Dr. Dudal went on, 'If all the land around a city is fertile farm land or productive grazing or forest tracts some of it has to be lost to the needs of expanding populations. But the whole perimeter of a city should not be lost that way. Expansion of a city can be on one side only. Efforts of this kind are generally not being made now, especially in Europe.'

Mr. Erik Danfors, of the Royal Institute of Technology in Stockholm, told the same meeting that 'As man grows increasingly urban he grows increasingly thoughtless and careless of rural man and of rural life in general. City man is particularly unconscious of soil degradation and loss. He just does not think of it. Today the decision makers are increasingly men with little connection with nature.'

We must awaken an urban consciousness of the soil. With air, water and sunlight it is one of the basic resources of the planet, a fundamental biological requirement to all land-based life. It is the urban-industrial attitude to food production that has led the agriculture of the developed countries into so many blind alleys. Factory industry is a linear process; a raw material is subjected to a series of processes ending with a commodity. It is tempting to regard agriculture as an industry in the same light and, in fact, it is so regarded today. But there is a fundamental difference. Agriculture is cyclical. Its raw materials are living organisms and its processes are biological processes, not physical or chemical alone. If we come to regard it as an industry we may find ourselves attempting to maximise production to a point where we are 'mining' the fertility of the soil. In industry there is, in theory at any rate, no limit to the extent to which we can increase the throughput of the raw material, beyond the final exhaustion of that raw material. This exhaustion must be inevitable if the raw material is non-renewable. In agriculture, the raw material is renewable and the continuation of agriculture depends upon its being allowed to renew itself. If we 'mine' the soil we are exhausting a resource more basic than any mineral. The exhaustion of minerals might spell the end of industry. The exhaustion of the soil would spell the end of life. It is of some importance that we subject ourselves to a psychological 'de-urbanization.'

Urbanization and industrialization are not the only predators of land. There is a buffer zone between the fertile land of many of the developing countries and the desert. These 'marginal' lands must be stabilized. It is not a question of bringing them into production; they must be fortified to resist the encroachments of the desert. Indeed, there are grave dangers in bringing them into agricultural production. In some places this has been tried. New water boreholes were drilled on the southern edge of the Sahara which made cattle ranging possible. This encouraged overstocking. Patches of bare soil are showing, streams are drying up and the sand dunes are forming and spreading. Deserts are often surrounded by areas of scrub land, brush or forest, and ecologically they may be highly unstable. If they begin to deteriorate, they may reach a point at which, quite suddenly, they collapse. Erosion may accelerate, accompanied and aggravated by flash floods, leading to dust bowls, spreading quickly over wide areas. The most likely means of stabilizing such soils is through afforestation, but this costs money.

There are other, more exotic ways in which the ecology of some underdeveloped regions is being changed. In 1963, 600 people died in a village called San Ramon, in the Beni State of Bolivia, deep in the forest. The Bolivian Government called in the American Institute for Tropical Diseases and it took two years to diagnose the cause of the deaths as haemorrhagic fever, a rare and dangerous disease which is widespread in small mammals. Upon further investigation it was found that there was an unusually high number of rodents in and around the village. This could be accounted for only by the fact that there had been a decrease in the population of the rodents' natural predators – jaguars, wildcats, pumas and ocelots. The predators had been killed for their skins. The hunting of these animals is illegal in Bolivia, but the country is large and the jungle dense, and in it there are numerous small airstrips which it is impossible to police. The temptation to the villagers is large. Although the price they are paid is an insignificant part of the price of the finished garment, it is wealth to them. Yet, by undermining the ecological stability of the region the villagers are killing themselves. Haemorrhagic fever broke out again in early 1971.

In Sierra Leone the small, grey, vervet monkey is multiplying as its natural enemy, the leopard, disappears. Now the monkeys are invading cultivated fields and destroying crops. Yet it is, in part, the creation of the fields that has caused the problem. The leopard's habitat has been reduced.

Ecosystems may continue in a state of complete stability for very long periods, but in spite of this their stability is maintained by an intricate network of relationships that is easily disrupted.

There is a further danger in the removal of areas of wilderness. All of our crop plants and domestic animals have been selected from wild species and then bred for particular characteristics that man favoured. We cannot be certain that our present tastes will endure for all time. Indeed, agricultural scientists and ecologists are searching all the time for new species that might be cultivated and the time may come, and soon, when it will be seen to be highly desirable to seek new plants and animals that are more productive than those we cultivate at present, or that have a higher nutritional value, or that can be crossed with existing varieties to give improved characteristics. If we and our descendants are to have the capacity to make such changes, then we must permit the wild species to survive, not just as museum specimens or as seeds, but as plant and animal communities, viable and self-sustaining. The earth's wild flora and fauna constitute a genetic pool, a bank if you like, on which we may wish to draw. As more and more land is cleared for agriculture or for urban or industrial development, this pool is depleted. We are destroying one of the planet's most basic and irreplaceable resources and committing our descendants to those foods that we have developed and prefer. The most charitable view is that this is the height of irresponsibility. A more realistic, if harsher view might be that in order to satisfy our immediate needs in the easiest and most convenient way we are prepared to lay down immutable conditions that will affect the lives of all who come after us and to throw away a freedom of choice and action that has been part of man's heritage since the beginning of his tenancy of the planet. Fortunately, steps are being taken to check this destruction, but without the support of governments it is difficult to see how the efforts of scientists alone can halt 'the march of progress' towards urbanization and industrialization. A seminar on 'Plant exploration, introduction, maintenance and utilization of the genetic resources of cultivated plants' was held in June 1971, in Moscow, Leningrad and Krasnodar. It was sponsored by the FAO and the government of the USSR. To some people conservationists – those who would preserve areas of wilderness – appear romantic, sentimental and, perhaps, as representatives of a middle class anxious to preserve its privilege of solitude. In view of the hunger that exists in the world and the need for land, it is easy to think of them as selfish or to dismiss them as cranks, out of touch with reality. In fact, they are fighting a desperate rearguard action to protect an asset that is of importance to all of us, even though we may not be aware of it.

This is of direct relevance to the world food problem. Any attempt to extend agriculture into new lands may bring disproportionate environmental disadvantages. Dr. M. L. Rosenzweig, of the State University of New York, has warned of dangers inherent in our efforts to increase agricultural production. Basing his predictions on calculations made from mathematical models of variations in populations of predators and their prey, Dr. Rosenzweig has said that by increasing crop yields we may increase the populations of herbivores to such an extent that the carnivores would be unable to keep them in check. The result could be a net reduction in food output. This is a controversial assessment and many biologists are extremely cautious about accepting mathematical evaluations of complex ecological relationships. Nevertheless, eutrophication, which leads to a proliferation of microflora, also kills the fish that eat microflora as they cannot breed fast enough to cope with the sudden increase in food supply and lose the competition for oxygen.

Warnings of the risks inherent in increasing agricultural production are legion. Every day there are examples of damage to the environment, often to people as well, by agrochemicals, especially pesticides. In Britain, warnings of the deterioration of the structure of the soil are reported frequently. Are they heeded? In the developing countries, too, there are fears of the ecological effects of human expansion. On the whole, they are being resisted manfully. Mr. T. N. Sarif, India's Permanent Representative to the FAO, told a preliminary meeting for the UN Stockholm Conference that 'we will resist attempts to place a disproportionate emphasis on the problems of human environment if they make us lose sight of the needs for further economic development of nations of the third world.'

There we have it, in what must be one of the most extraordinary statements in history. Industrial and economic growth must not be held back. They must be allowed to continue indefinitely, no matter what effect they have on the quality of the lives of the people they are meant to benefit! People must not starve. If they are poisoned by pesticides, that is too bad, but at least they did not starve. People must be wealthier. What does it matter if they live in overcrowded cities, breathe foul air, drink tainted water, eat adulterated and devitalized food for which they pay exorbitant prices, so long as they have money? It is staggering, but, alas, it reflects current economic theory in which all activity is equally good, so that the manufacture of the gun, the consumption of the round of ammunition, the transportation of the victim to the cemetery and the gunman to jail are all adding to the total activity of the community and so are equally good. We may, perhaps, understand the attitude of the man who says 'What has posterity ever done for me?' but at this the mind boggles!

So what of the world? Who will eat? Who is eating now? In Senegal they are making bread from a blend of wheat and millet flour. They hope to save 600,000 dollars in wheat imports in the first year. At present, the world produces 330 million tons of wheat a year and it is not enough. In 1968, South America imported 365 million dollars worth, Asia 1,660 million, Africa 407 million. Peru alone spent 50 million dollars on the import of 600,000 tons, China imported more than 5 million tons, at a cost of 338 million dollars. Still it is insufficient to go round. So the developing countries, where people are acquiring a taste for bread, are exploring the possibilities of diluting wheat flour with other flours made from manioc, groundnut, soya or millet.

Attention is also being given to the possibilities of harvesting wild game. This is sensible, because indigenous animals are better adapted to local conditions than stock imported originally from temperate countries.

Meanwhile, in the developed countries, the development of new crop varieties is a highly profitable business, profitable enough to attract industrial spies. A company in Colorado, which believes it may be about to launch a new wheat that will increase yields by 25 per cent, locks plans of its experimental plots in a safe. The project is attracting heavy investment. In spite of the enthusiasm, however, it is unlikely that even if it succeeds, the new hybrid will benefit many farmers in the long term. It will demand high inputs of agrochemicals. Immediately, the profits may be huge, as American farmers adopt it, but it is doubtful whether the yields it promises can be sustained, and it is certain that its contribution to the world food problem will be minimal.

In Canada the farmers over-produce and the government advises maximum acreages that should be sown. The surplus is stored and as a result of the reduced crops in 1970, the stocks have fallen. This is satisfactory for the Canadian farmers, but it is the world's reserves of grain that are called on in emergencies. In 1971 the USSR had to import wheat because of a poor harvest from the winter wheat planted late in 1970. World wheat stocks have been reduced by 27 per cent.

Over-production is as much of a problem as under-production and, because of the ways in which international trade operates, surplus food is not necessarily available to those who need it. In 1970 the rice growing countries of south east Asia had a good harvest, and the FAO had to urge the developed countries that export rice to restrict trade in order to give the developing countries access to world markets.

As a result of reduced production in several of the developed countries and a major disease – southern corn leaf blight – that has swept across the United States, reducing drastically the maize crop of the world's third largest maize growing country, grains are in short supply throughout the world as a whole. As developing countries are forced to import food, the trade benefits farmers in the developed countries. The rich grow richer again! In 1969 world trade in agricultural commodities increased by 4 per cent – the first increase for three years. Two-thirds of the increased income went to the developed countries. Part of the growth was due to reduced output in importing countries, coupled with increased domestic demand. Total world agricultural production in 1969 remained static, and where there were increases in exports from developing countries, they were mainly in non-food products, so that such exports barely compensated for food imports. In Africa especially, food production is declining because of falls in cereal production.

Apart from North America, where production is either held back or reduced by disease, food production in the developed countries is static, or falling, and has been for several years. As we saw in our consideration of the present situation and likely future for British agriculture, no major increases are likely.

In the developing countries, as we have seen, the situation is no better. In 1970 there was a poor harvest in the Near East, due, according to the FAO, to bad weather. Production remained static, while the population increased by 2.8 per cent. The wheat harvest was down by 5 per cent due mainly to declines in Afghanistan, Iraq, Jordan, Syria and Turkey, but the barley yield was down by more than 20 per cent. Maize and rice yields remained at their 1969 levels. In Africa, production in 1970 rose by 2 per cent, but population increased by 2.7 per cent. The pattern is repeated throughout all the developing regions, except the Far East, where there was a good rice harvest – which there was difficulty in selling because of competition from the developed countries. Even in this region, however, the improvement was not spread evenly. Taiwan, and Sabah and Sarawak (in Malaysia) showed no increase and the increase in Ceylon, Laos, Pakistan, Korea and Thailand amounted to no more than 2 per cent or less. The real gains were made in Khmer (25 per cent), Vietnam (8 per cent) and the Philippines (7 per cent). There was a 5 per cent increase in India, due mainly to increased yields of rice and wheat.

Dr. E. M. Ojala, Assistant Director-General in charge of FAO's Economic and Social Department said in Rome on 11 November 1971, that, 'The growth in output for the developing regions as a whole ... would allow for no increase at all in *per caput* terms in 1971.' He praised the achievements of the 'Green Revolution,' but his praise was qualified:

'But the fabric of this progress in the developing world as a whole remains fragile and its base narrow, both geographically and in terms of commodities. ... In 1971, India again had a good year, but there were setbacks in Ceylon, Pakistan, and the Philippines, and the latter country has again had to import rice, after being self-sufficient since 1967.

'For the first time in the short history of the high-yielding varieties programme, an important reason for the setback at the national level, in the Philippines, was a major outbreak of disease among the new rice varieties, an eventuality that had been feared from the start.'

The immediate future appears gloomy. Increases in production are expected, but they will come no where near meeting the IWP requirements by 1985. Food production and population growth are engaged in a race in which population growth maintains a steady pace. Each time food production falls behind it must move that much faster to catch up again, and, it seems, the pace that was calculated in the first place, the rate of increase that would meet the IWP 1985 target, was too slow and the target too low. On 29 September 1971, the FAO published its *Agricultural Commodity Projections, 1970–80*, in which it projects an increase of 26 per cent in food production by 1980, mainly in the developing countries. This is an annual increase throughout the world as a whole of an average 2.5 per cent a year, linear, not exponential growth.

'The growth of world food and feed production is projected to slow down moderately to an annual average of 2.5 per cent in the seventies from 2.8 per cent in the period 1959–69.'

The report is guardedly optimistic, nevertheless. It maintains that the slowing of the rate of growth will be due to reductions in the developed countries, but that in the developing countries production will increase faster – slightly faster – than demand, and it has high hopes of the 'Green Revolution,' which it believes will lead to 'a sustained and historically high rate of growth of cereal production in developing countries.'

It is difficult to find justification for the optimism or, indeed, for any optimism at all. Even if the projections are accurate, and the introduction of the new varieties were to be successful and if all the necessary inputs for them become miraculously available, the best that can be hoped for is that by 1980 the situation will be no worse than it was in 1970. Dr. Boerma, in his foreword to the study, says:

'Nobody can contemplate without dismay the finding that even by 1980 the absolute number of people short of food may be much the same as today even though the average shortfall should be somewhat less.'

Even in countries where, statistically, enough food is available, inefficient distribution and wastage often leaves many of the poorer sections of the population underfed. Production did increase during the 1960s but, according to the *Projections*, 'as world population increased during the decade by 680 million inhabitants while the average dietary level improved only slightly, it is possible that in absolute terms the numbers of undernourished people actually rose and that nutritional problems were therefore no less acute at the end than at the beginning of the decade.' Even if the income levels that are called for by the FAO were attained, there are forty-two countries in which the total calorie requirement for altogether 1,440 million people will not be met. The *Projections* concludes that there will be a need for food aid at least as great, and possibly greater, than there was in 1970. At the present time, as we saw, world grain stocks are reduced.

Mr. Boerma gave an address to the 18th General Conference of the International Federation of Agricultural Producers in May 1971. In it he said:

'All in all, throughout the world, not an encouraging picture. It would of course be distorting things if I failed to mention any of the great positive achievements that have occurred in world agriculture in the last twenty-five years. The most obvious, perhaps, is the "Green Revolution." But, even here, I must urge the strongest caution in our expectations. There is no doubt that the development and introduction of the high-yielding varieties are a technological breakthrough of the greatest importance, holding out hopes that we may finally be able to oust the spectre of famine from this world. But, quite apart from the vast material and policy backing that these varieties will require, it is no play of metaphor to say that they contain in themselves the seeds of new dangers. We have already found that, due to their greater immediate potential in irrigated areas rather than dry or rainfed ones and by benefiting rich farmers while leaving the poorer ones no better off than before, they could not only help to perpetuate, but even aggravate, existing social inequalities. Furthermore, the increased production to which they are leading is almost certain, in the absence of adjustment policies, to add further complications to international trade in cereals.

'If what I have been saying sounds gloomy, that is exactly what it is meant to do. For I believe that the countries of the world must face up clearly to the implications of where their policies in matters of agricultural production and trade are leading them. And, inasmuch as these implications are potentially disastrous, they must also see the reasons *why* their present policies have generally gone wrong.'

If there is to be an improvement, it must come from the rich countries. There must be a major redistribution of wealth. With this, and with an understanding of the nature of the problems and a real will to solve them, there is a possibility that a little more time may be bought in which the world's population can be brought to a level at which it can be fed from the planet's renewable resources.

The alternative, as Mr. Boerma said, is disaster. It is not merely that strange people in far-away countries will continue to grow hungry. It is far worse than that.

By the late 1970s or early 1980s, the demand for food in developing countries may exert such pressure on the governments of those countries that they will have no choice but to restrict exports of food – those that do export – and divert such food as they have to their domestic markets. They may be compelled to channel agricultural production from cash crops – cotton, tea, coffee, cocoa, rubber, etc. – into food production. This will lead to immediate reductions in imports of food and certain raw materials into the developed countries which, in turn, will make them more immediately dependent on what they are able to produce themselves. This will deplete still further the world food reserves as surpluses, particularly of grain, are diverted to animal feed. The pressure to intensify production in developing countries is likely to accelerate problems of erosion and soil depletion and as areas of inferior land are forced out of production the pressure on what remains will increase still further. Reductions in export earnings will make it more difficult for these countries to buy in the technological materials and service their agricultural systems are being taught to need. Major famines may occur in Latin America, the Near East and Africa and the world will not possess stocks for emergency use to deal with them. The famines may be accompanied by epidemic disease and this may spread into Europe and North America.

By the mid or late 1980s the depletion of petroleum reserves will necessitate expensive re-equipping of the farms in the developed countries with machines that derive their power from alternative sources – if there are alternative sources by that time. The price of agrochemicals will rise steeply and then, before the end of the century, supplies may fail completely. This will cause substantial reductions in production in the developed countries unless alternate systems of agriculture have been developed and introduced that are geared for sustainable yields in the conditions that will obtain by the 21st century. In the absence of such systems, there may be famine in Europe and North America, again accompanied by epidemic disease.

As resources dwindle there may be wars between developing countries for control of what remains and as the gulf between rich and poor nations continues to widen tensions between the two groups will increase. Since the industrial countries are heavily dependent on imports of raw materials from the developing countries, any serious attempts by these countries to divert resources to their own use – including food resources – may provoke military intervention in their affairs from the developed countries. This would increase tension between the rich countries, which are competing already for what reserves remain. It is not impossible that these tensions might erupt into global war between the major powers.

That, then, is the picture. There may be increasing social disorder throughout the world. In the developing countries civil wars, revolutions and wars may accompany famine and disease affecting whole regions. There may be wars between rich and poor nations, famine and disease in Europe and North America and, eventually, global war.

The warnings are all there now. The hunger is there, the discontent, the social instability. Year by year the rich grow richer by depriving the poor of their share of what the world can offer. Populations grow and our governments have yet to learn that continuous and unlimited expansion – of anything – is impossible within a finite system, and that this planet is finite.

Disaster is not inevitable, even now, but will it be averted? There is little sign at present that the extent and nature of the crisis has penetrated the consciousness of governments, far less stimulated them into taking action. They continue in their naive trust in technology and its omnipotent ability to solve the problems it creates. They are awaiting the '*deus ex machina.*'

There is no natural law to protect us – we have violated natural law. If this civilization is on the verge of sudden collapse, history provides many precedents. We are not the first to outrun our resources. Malthus has been encountered before and the knowledge of imminent catastrophe has never implied that the community was bound to take action to avert it. No one listened to Noah!

There is not much time, perhaps five years, at the most ten. The ball is in our court. The world food problem is our problem. It is here and it is now. It exists in the boutiques of London, the factories of Birmingham, it is in New York and Paris, Washington and Moscow. It is there that it began and it is there that it must be solved. It is for us to make ourselves aware of the world we live in and to force our awareness on our governments. It is our world. Ours and our children's. We must save it if it is to be saved. There is no one else.

INDEX